FOR A NEW GEOGRAPHY

FOR A NEW
GEOGRAPHY

Milton Santos

Translated and with an Introduction by
Archie Davies

UNIVERSITY OF MINNESOTA PRESS
MINNEAPOLIS • LONDON

I dedicate this book to the memory of a geographer
of open heart and mind

LYGIA FERRARO

She fought for a more generous geography, a new geography.

Contents

Translator's Introduction

The Newness of Geography

Archie Davies

For a New Geography signaled the emergence of Milton Santos as a major interpreter of geographical thought and one of the preeminent global theorists of space. Published in 1978, it was not his first book—a number had appeared in both French and Portuguese, and he was a well-established figure in disciplinary geography—but it set a new high-water mark for the ambition of his intellectual project. It both signaled and helped to produce a moment of transition in the history of geography. In that same year, for instance, also appeared Derek Gregory's *Ideology, Science and Human Geography,* and the book came out just as the critique of quantitative geography was crystallizing and a new critical geography was being consolidated (Barnes 2017). It is strange, therefore, that it is only now appearing in English. It was first published at a moment of transition, but it functioned, also, as a bridge; in moving forward to his own vision of a geography of liberation, Milton Santos began by turning back to the roots of modern geography itself. It was published in French in 1985 as *Pour une géographie nouvelle,* and it appears now at another propitious moment, when geography is reassessing its critical past. There is a renewed interest not only in alternative geographical traditions but also in the institutional history of radical geography itself (Ferretti 2019; Theodore et al. 2019). This book serves both tendencies: it occupies a distinctive, dual place in the history of geography, being at once a critique of the Northern, Anglo-centric discipline from within and a systematic critique of its flaws and assumptions from outside. That it is being published in English only some forty years after its publication is an ineloquent reminder of its continuing relevance.

The new critical geography of the late 1970s was informed in the Anglophone world by rereadings of Marx and an engagement with the socioenvironmental context of underdevelopment. This latter context—and the literatures and politics it might presuppose of anticolonialism, dependency theory, and Third Worldism—was, however, to remain underdetermined and neglected in Northern, Anglophone geographies (Power and Sidaway 2004). Though many exceptions could be posited, the upsurge of scholarship in the fields of environmental justice and postcolonial geography in the 1990s and 2000s and, more recently, in subaltern, Black, and decolonial geographies (McKittrick 2006; Jazeel 2014; Daigle and Ramírez 2019), coupled with an expanding critique within scholarship on the history of geography of its Northern, white, male hegemonic tendencies (Maddrell 2008; Esson et al. 2017; Craggs and Neate 2019), show that things have been missing in the critical geography that emerged in the late 1970s and 1980s. There is no panacea for a lost critical geography in Milton Santos, but his unique set of concerns and his own intellectual and personal biography can disrupt and destabilize the concepts of centrality and the canon in the history of the discipline.

Milton Santos was born in northeast Brazil, in Bahia, in 1926. He came from a middle-class Afro-Brazilian family and both his parents were teachers. He studied law at the University of Bahia in Salvador, taught by major Brazilian intellectuals such as Luís Vianna Filho, before becoming a geography teacher (Contel 2014, 939). He then worked as a journalist, largely for the newspaper *A Tarde*, writing about left-wing politics, regional underdevelopment, and the cocoa region of Bahia. In becoming editor of the newspaper, he secured himself a place in public life (Contel 2014, 940). Between 1956 and 1958 Santos wrote his doctorate in France, under the supervision of Jean Tricart at the University of Strasbourg. He returned to Brazil to work at the University of Bahia as a journalist but quickly became heavily involved in politics. During the early 1960s, northeast Brazil was a hotbed of political thinking and activity, and Santos—and his brother, Nailton—were both involved in regional politics, marked by emergent radical visions of national and regional economic planning and alignment with Third Worldist liberation movements (J. de Castro 1966; Kirkendall 2004). As a young, radical bureaucrat Santos traveled to both Cuba and newly independent Africa in this moment of tricontinentalism (Moreira 2010; Contel 2014). He was appointed president of the Economic Planning Commission of Bahia by then Governor Lomanto Júnior. With the military coup of 1964, however, Santos was targeted, arrested, and imprisoned. He was ultimately forced to leave Brazil. He went to France, initially living and working in Toulouse and later in Bordeaux and in Paris. While in Paris he was an important part of the French national geographical debate (Pedrosa 2015a, 17–19; Ferretti and Viotto Pedrosa 2018), including editing two special issues of *Revue Tiers Monde* and working for the UN as an adviser in Venezuela (Contel 2014, 944–46; Vasconcelos 2018, 48). His was an itinerant academic's life. He worked for a stint at the University of Toronto and was invited to teach at University College London. In an ignominious turn for British universities and for academic geography, he described arriving in the city, receiving no advice or support, and leaving shortly before taking up his post after being racially abused and discriminated against while seeking housing (Vasconcelos 2018, 49).

The second half of the 1970s were an important period for Santos, not least because this was when he wrote this book. He worked—as the dedication makes clear—in a number of different universities, including at the University of Dar Es Salaam between 1974 and 1976, a period of intense intellectual activity in the country under the socialist government of Julius Nyerere. He was one of a number of absorbing figures who passed through Dar at this time, including Walter Rodney and others, when the university was a hub of intellectual radicalism and the city a nerve center for anticolonial movements across East Africa, from FRELIMO in Mozambique to ZANU-PF in Zimbabwe and the antiapartheid movement in South Africa (Sharp 2019). The geography department itself played host to significant figures in Anglophone radical geography, including David Slater and Joe Doherty, both of whom were interlocutors with Santos both then and afterward, and Ben Wisner, an important early writer of political ecology who passed through the department frequently while working in Kenya. While there, Santos traveled extensively across East Africa—often with his son, Milton—and worked not only on his important book on urban geography but on this book. He then went to Caracas, to the Central University of Venezuela. Both of these experiences were clearly influential, as can be read not least in the reference list to this book. It includes the work of colleagues from both places and a number of manuscripts and documents—by Roberto Briceño, Sonia Barrios, and others—that are very difficult or impossible to access. During his peripatetic period, Santos consolidated an international network of critical scholars (Ferretti and Viotto Pedrosa 2018), and when he returned to Brazil in 1977 he embedded himself in the country's intellectual life.

In the period in which he was writing *For a New Geography,* Santos was taking an active part in global networks of geographers who were beginning to challenge not only geography's long-standing traditions but its quantitative turn. More important, a group of critical geographers was beginning to ask what the role of space was in the reproduction of capitalist social relations and what Marxist and geographical theory and praxis might have to say to one another. Santos's role in this broad collective was vital: he brought to these debates a focus on underdevelopment and Third Worldism. A few years before the publication of *For a New Geography* in Portuguese, one of Santos's works, *Shared Space* (1975), was published in English. That book positioned Santos as an urban geographer with an economic bent. It laid out his theory of the double circuit of the economy in urban space. It was influential among urban development geographers and made him a figure in Anglophone geography (Contel 2014, 946–47). As laid out in a 2017 symposium published by *Antipode,* the journal of radical geography, Santos edited a number of that same journal in 1977 and published six important essays in it between 1974 and 1980. That work was cited by English-language scholars at the time and since as being key to the emergence of critical geography (Peet 1985; Featherstone 2019). He also published a little-commented-on series of lectures, *Underdevelopment and Poverty: A Geographer's View,* through the University of Toronto in 1975, and some hard-to-access articles and chapters on the social economy of cities (Gappert and Rose 1975), growth poles (Santos 1975), and central place theory (Santos 1979, 125; Santos writes that the essay, published as chapter 4 of *Economia espacial,* was first published in English in Paris in 1973, though I have not been able to locate a copy). After 1980, his engagement with Anglophone geographical debates took a downturn as he increasingly focused on the Brazilian context (Ferretti and Viotto Pedrosa 2018). This was an intellectual choice; Santos believed that for a country to move out of dependency it needed to produce its own intellectual tradition, and he wanted to be part of this for Brazilian geography (Santos and Bernardes 1999; Contel 2014). He was invited to write for *Antipode* a number of times by Joe Doherty, the editor of the journal between 1986 and 1992, but this invitation never came to fruition (Doherty 2020). In 1983 he secured a permanent position at the University of São Paulo, which was to become his intellectual home for the next two decades. Throughout the 1980s Santos published works that consolidated his position as the most influential geographer in Brazil; many of these works can be seen as leading toward *A natureza do espaço: técnica e tempo; razão e emoção* [The Nature of Space: Technics and Time; Reason and Emotion], in 1996. Based in São Paulo, he became a public intellectual and a deeply influential force on his discipline in his home country. He continued to write about urbanization, globalization, and geographical theory until his death, in 2001.

There is not one Milton Santos in the history of geography but many. There is the seminal, absolutely central—perhaps even overbearing— figure of Milton Santos in Brazilian geography (for an account of the debates in critical geography in Brazil see Pedrosa 2015a, 2015b). For geographers of a critical bent (and indeed for those without) in Brazil in the second half of the twentieth century and the beginning of the twenty-first century, Santos and his oeuvre were unavoidable points of reference. His work is published in dozens of editions; his most far-flung and incidental writings are gathered into closely edited collections; his work is analyzed in dozens of volumes of essays and features on the reference list of thousands of journal articles. Then there is the Milton Santos of the global Francophone and Hispanophone circuits of geographical thinking (as well as Italian, Catalan, and others). There, Santos is a major figure, works such as *A natureza do espaço* having long ago been translated into Spanish and French, and a significant part of his writing first appeared in French. Dozens of conferences and books have analyzed and deployed his work. Finally, there is the English-language Milton Santos, a much more fragmented, uncertain, and emergent figure. This translation is part of a disparate collective effort to fill out and fill in Santos's Anglophone reputation, to make known his unique contribution to the history of ideas about space, nature, globalization, territory, and urbanity in the twentieth century. It aims to reemphasize not only that the history of geography has always been multilingual but that it has long had an anticolonial and anti-imperial tradition within it. For those who want to decolonize geography, one way to do it is to look more closely at the discipline's own history, particularly beyond the English language.

Santos's major works have remained untranslated until recent years. Now a number are beginning to appear, including *Toward an Other Globalization: From the Single Thought to the Universal Conscience* (2018). A companion volume, *Milton Santos: A Pioneer in Critical Geography from the Global South*, explores how his work might be put to use in Anglophone geography (Melgaço and Prouse 2017), and a symposium in the journal *Antipode* has helped (re)introduce Santos to the English-speaking geographical community. This has coincided with fresh attention, from scholars such as Breno Viotto Pedrosa and Federico Ferretti and from the Rede Brasilis network, to the histories of critical Brazilian geography and their intersections and interactions not only with global flows of knowledge but also with contemporary debates over the geopolitics of knowledge, space, nature, and uneven development. This context makes it a particularly propitious moment for this book to appear in English for the first time.

For a New Geography

For a New Geography is a comprehensive introduction to Santos's work. Its project is captured by the direct and lucid title; he believes that geography needs renewal, through a recommitment to ontological space. But Santos's original subtitle is important: *From a Critique of Geography to a Critical Geography*. In this edition we have reinterpreted this subtitle to name Santos's introduction (in the original called *Introdução*). We have done so to avoid the triple repetition of "geography" in the title but also to emphasize the simplicity of Santos's central project: *For a New Geography*. It is crucial, nevertheless, that Santos is a deep reader of the geographical tradition—what is new in his geography emerges, in part, from the old. It is far from his final statement on these issues, but it systematically lays out the scope of his lifelong concern with defining the object of geography, space. It appeared at a crucial moment in Brazilian geography. In 1978, with Santos having just returned to Brazil, the publication of the book coincided with a conference of the Association of Brazilian Geographers in Fortaleza. It was his first major intervention after returning from his effective exile under the military dictatorship. His appearance in Fortaleza and the book were highly influential in driving a leftward movement in Brazilian geography (Grimm 2011, 177; Ferretti and Viotto Pedrosa 2018). This is a vital context for the book, which was part of a conscious project to build a Brazilian critical and radical geography. As such, the book is itself a project of cultural translation. It coincided with the loosening of the Brazilian dictatorship and the beginning of a decisive push toward democracy. In disciplinary terms, then, the book was integral to a political shift, although the book's own politics are almost entirely intellectual; nowhere does Santos make arguments about contemporary political situations, and the language of his essentially anticolonial, Marxist Third Worldism is restrained and implicit. Indeed, one feature of the book's particularity is the persistent quietness of its author in its pages. It is very self-conscious of its own geography—this is clear, not least, in the acknowledgments—but nevertheless sets out to proclaim a set of abstract, total theorizations to which Santos's own position as a Black Brazilian is apparently incidental.

In the Introduction Santos lays out that *For a New Geography* will be the first of a projected series of five works under the "general theme of *Human Space*." These five books, he writes, will also include *From Cosmic Nature to the International Division of Labor, Spatial Organization and Contemporary Society, Social Time and Human Space,* and *Social Totality and Total Space: Form, Function, Process and Structure.* In fact, Santos did not write books with any of these titles. This is not to say, however, that he abandoned the project. On the contrary, he continued to elaborate and develop these concepts throughout his career. He did publish one book, *Metamorphoses do espaço habitado: fundamentos teóricos e metodológicos da geografia* [Metamorphoses of Inhabited Space: Theoretical and Methodological Foundations of Geography] (1988), which was specifically a sequel to *For a New Geography,* updated for the conditions of Brazil in the late 1980s. However, the continuation of *For a New Geography* can also be found across his writing, in works such as *O espaço do cidadão* [The Space of the Citizen] (1987), *Técnica, espaço e tempo* [Technics, Space, and Time] (1994), and *A natureza do espaço* [The Nature of Space] (1996). Across his career his ideas changed, and it is easy to overstate the coherence of his theoretical production, but the same concerns that he identifies in the introduction—the spatial organization of society, the relations between time and space, the dialectics of space, modes of spatial distribution— are all explored in many of his books and essays.

For a New Geography displays Santos's characteristic rigor and sys- tematism. He begins by placing himself squarely in the center of the his- tory and present of disciplinary geography. Santos orients himself toward the French tradition—the book's opening paragraph references Emman- uel de Martonne and Jean Brunhes—and it continues to be the lodestar of his enterprise. This is unsurprising; French and Brazilian geography have always been bedfellows, and Santos was trained and practiced in a transatlantic space in which the European was always predominately understood through the French. It is possible to trace in Santos's refer- encing practices the circles in which his thought moved. This referenc- ing, too, positions the metadisciplinary debate that the book engages in as central to the trajectory of geography itself. To read the book is to be deeply immersed in the big debates of geographical theory of the late 1970s. His reading, we should note, was not just in Anglo-Saxon and French literatures but also across Nigerian, Italian, Soviet, and other Latin Amer- ican geographers.

In the opening chapters Santos outlines precisely why he sees geog- raphy as in need of renewal. The world, he argues, has changed in a revolutionary way, and so science and knowledge must change with it. Geography has been overcommitted to an underdetermined set of theo- retical premises. A new geography is needed because old geography— official geography—has failed to grapple with its own object: space. To confront this "grave epistemological error," Santos will provide not only a new theory of geography but also a new epistemology. He will not do so alone, and the opening chapter reveals not only the breadth of the work that he envisages but also the wide sphere of reference that he will draw on—from Darcy Ribeiro to Jean-Paul Sartre and Bertrand Russell. His intellectual influences are, it is important to note, dominated by French writers, including work such as Lucien Febvre's *La Terre et l'évolution humaine*, which is perhaps more influential on Santos's thinking than the direct citations suggest. The role of French geographers following Vidal de la Blache is also critical. The range of his referencing suggests the masculine dominance of Santos's intellectual topography. Though he worked with and cited many women throughout his career—notably important Brazilian geographers such as Maria Adélia Aparecida de Souza—he, like the majority of male radical geographers of the period, was late and myopic when it came to the need for a feminist critique of space and geographic method.

After a sweeping introduction, Santos turns back to unpack the development and, inevitably, the failings of geographical research. In a rigorous, often painstaking manner, Santos works his way through the qualities and flaws of the geographical tradition, beginning with what he calls the "founders" of the late nineteenth century through the late 1970s. This is a thematic, not a chronological, critique. He analyzes geography's ideological bent and its roots in colonialism, imperialism, and European expansionism. He traces the dominance and decline of determinist approaches and the emergence of the Vidalian school and Cultural Geography on both sides of the Atlantic. Here he necessarily confronts the concept of the region. Alongside topics such as technics, globalization, and underdevelopment, the region was a recurring theme of Brazilian debates in the social sciences from the 1950s to the 2000s, and one to which Santos would return again and again in his later work. It is also a field in which contemporary geographers have much to gain from reading Santos's work (see for instance I. E. de Castro 2002; Santos 2008, 2009). While he here declares the "liquidation of traditional regional geography," this is a liquidation of a particular sort: one that recongeals in new and unpredictable forms and in which the concept of the region remains a highly active solution.

In the opening chapter, Santos reflects on geographical epistemology and challenges both the uses of analogy and the epistemological relations between the physical and social sciences. This leads to a meditation on the philosophical sources of geography (chapter 2), pulling apart the different inheritances of Descartes, Kant, Darwin, Comte, Hegel, and Marx and how they variously influenced diverse fields of geographical research.

Overall, Santos argues, geography's philosophical eclecticism has compromised its epistemological coherence.

Santos turns next to the postwar history of geography and a thoroughgoing critique of both traditionalist ways of proceeding and the quantitative revolution (chapters 3–7). Writing in the late 1970s, Santos was treading on a freshly ploughed field, and, rereading his work now, we can sense that the intellectual stakes were high. While Santos's critique of linearity and colinearity, regression models, systems analysis, behaviorism, and other tools of quantitative analysis are not, perhaps, topical in the twenty-first century, it remains fruitful for at least three reasons. Firstly, it constitutes an exemplary critical history of twentieth-century geography up to the late 1970s. Second, there is a real historical value to be found in closely reading this critique of quantitative geography from the South. In the context of ongoing debates over decolonizing geographical ideas, this dismantling of quantitative geography has an important place: an assessment both from inside the scholarly geographical world and from a Third Worldist perspective. Finally, Santos's core arguments remain relevant. His critique of empiricism, the excessive power of measurement, and the ideological risk that quantitative approaches reorganize reality in their own image can all be laid at the door of parts of contemporary social science and the powerful new technologies that have taken on dominant positions in academic enquiry inside geography and beyond. At the end of the first part of the book, this history of the discipline arrives at the late 1970s to argue that geography is in crisis (chapter 8). Among other things, Santos argues, a new modality of interdisciplinarity is required that places stringent demands on critical geographers (chapter 9). Allied to such new methodologies, though, comes the need to define space (chapters 10–13).

In part 2, Santos moves from critique toward a redefinition of the object of geography. Many complexities emerge here, as space is both subdivided—a permanent space and the space of our time—and redefined in the totality. He moves toward what would become familiar Santos terms for understanding space, but his definitions are never fixed but mobile and dialectic. Definitions of space become working rubrics for the many angles from which Santos approaches the object of geography. Among other things, he variously writes that space is a set of systems of actions, "socialized nature" (chapter 1), "a social fact" (chapter 11), "social order" (chapter 13), "worked matter," "landscape-space," "witness," and "an effective and active condition of the concrete realization of the modes of production and their moments" (chapter 12). When Santos speaks of defining space, he is not imagining a singular or straightforward endeavor. On the contrary, Santos's conceptual work, as part 2 of this book shows clearly, is at once a process of delimitation—he criticizes fallacies and rules out false categories—and generation. The function of definition is to effloresce, not to stultify.

By the end of part 2, Santos has laid out his critique of the history of geography and opened up a new set of definitions of space that attend to its *differentia specifica,* what he calls, citing Kusmin, "the specific logic of the specific thing." The scale of this task is clear: "all true theory is revolutionary theory" (chapter 14), Santos writes. He wants to construct a new paradigm, "a complete change in the vision of the world" (chapter 14). This complete change will be to address geographical theory to "total space in our time." Here Santos makes Marxist approaches—from Henri Lefebvre to Nikolai Bukharin—central, while drawing on a wide variety of influences. It is interesting, to note, for instance, the extent to which he draws on debates in Italian Marxism, in particular the journal *Critica Marxista,* and on French debates in the *Revue Tiers Monde.* Part 3 proceeds, therefore, to build upward from core concepts of the division of labor, transformations of social space, technologies, and forms of social cooperation across territories. His purpose is to parse the notion that to produce is to produce space. Yet, in a move familiar to much of his work, Santos moves from the abstract and transhistorical to the actual and concrete, analyzing the nature of space in the contemporary world. In particular, this includes a notable reading of the production of space—with particular emphasis on natural resources—in newly independent countries in the postcolonial conjuncture of the Cold War (chapter 15). His attention here to the dynamics of underdevelopment and "the geographical extension of the division of labor that now covers the earth" (chapter 15) yields an analysis both of scale and of the concept of *perverse universalization,* an early iteration of what he would later develop into a substantial research project on the spatial dynamics of globalization (see Santos 2017). Santos places significant emphasis on the state's role in the production of space (chapter 16) before analyzing what he refers to as the socioeconomic-spatial formation (chapter 17), which is a crucial tool in new geographical methodologies. The final chapter of the book introduces some further dimensions of this new critical geography: periodization, roughness, and relative time.

The Conclusion captures the final flourish of Santos's polemical tone. It places the need for a renovation of geography squarely within movements for social justice and global equality. He ends up in a position that is familiar to and that extends Marxist theories of space that were to dominate critical geography for the ensuing decades: "the reproduction of space emerges from the class struggle forged in the productive process itself" (Conclusion). Calling on the utopian socialism of Henri de Saint-Simon and Robert Owen, Santos yokes this New Geography to a broader project of liberation and the reconstitution of relations between humans and nature and between humans and their work.

Theory and Method

The book demonstrates two of the defining qualities of Santos's work across his career: first, a concern with conceptual elaboration—that is, with theory—and, second, a concern with the technics of doing geography—that is, with methodology. For Santos, the making of concepts is umbilically connected to methods of research and thinking. His critique of many of geography's existing traditions turns on his argument that they fail to articulate these two poles of intellectual work. Paradigmatically, here, we can point to the critique of quantitative geography that forms a major part of this book. He argues that quantitativists start from measurement and not from what is to be measured: "The great mistake of so-called 'quantitative geography' was to treat a method—indeed, a questionable method—as a theory" (chapter 4).

It is important to note that Santos's detailed concern with honing the methodology of geography is not an attempt to limit the discipline. Indeed, like his forebear Josué de Castro, he is explicitly driven by a vision of geography that is *abrangente*: broad, open and extensive (J. de Castro 1946). His concern with theory and method leads him to sustained discussions of interdisciplinarity. "New realities, requiring new explanations," he argues, "demand the appearance of new scientific disciplines" (chapter 9). He understands methodology as both politically engaged and historically contingent. It is through methodology that concepts become "operational"—a key notion for Santos that emphasizes his concern with geography that is put to work: what he would later characterize as "active" geography (see Grimm 2011; Bernardes et al. 2017). Elements of the discussions of interdisciplinarity in this book may feel somewhat dated to contemporary readers, but in this work Santos nevertheless raises important and unresolved questions about working across disciplines. Indeed, to some extent, as sciences such as physics and biochemistry become ever more technologized and complex, the possibilities for truly interdisciplinary work between natural and social sciences can seem more remote than ever. Yet perhaps there is much to be gained for geographers from reading Santos's enthusiastic engagement with physicists such as Einstein and contemporary theoretical science. Recent examples in the social sciences do of course exist—in and beyond geography—but they do not seem central to contemporary geographical theory-making as they are for Santos. For Santos, a dialogue with new scientific paradigms is crucial to New Geography, and the text is imbued with metaphors and ideas drawn from physics.

One of the book's theoretical interventions concerns the relationship between the totality, the socioeconomic-spatial formation, and the dialectical method. Santos's dialectical method was to develop much further in his later career, but we can register the concern here with process as a crucial waypoint of the direction his dialectical approach would take. This was a question not of a "dialectic of space. Or rather, a dialectic in space" (Introduction). For this dialectic to function, though, requires the concept of the totality. He draws his idea of the totality, at least in part, from the nineteenth-century European geographical tradition of Alexander von Humboldt, Carl Ritter, and Élisée Reclus. But he brings this into a fresh setting by establishing a direct dialogue with the place that he turns for a new methodology of geography: Marxism. "It is in the parallel between the creation of the means of production, the productive process, and the production and transformation of space that geographical method emerges" (chapter 15). The way of approaching this totality is through the idea of the socioeconomic-spatial formation. At the time, Santos was influenced by debates about Althusserian structuralism, which would later become less important to his work. How these concepts—of totality, socioeconomic-spatial formation, and the dialectical method—are interrelated becomes clear when Santos returns to the question of process in his elaboration of the concept of "geographic forms" in a journal article in *Antipode* in 1980. There he writes, citing Gramsci, that it is key to identify an "opposition between process and form":

> It is not sufficient to speak of process. Processes are nothing but an expression of totality, a manifestation of its energy in the form of movement; they are the instrument and vehicle of the totality's metamorphosis from universality to uniqueness. The concept of totality forms the basis for the interpretation of all objects and forces. (Santos 1980, 44)

Santos would return to the discussion of process in his magnum opus, *A natureza do espaço* (Santos 2009, 120–21) and in an essay on "form and time" published in 1992 (Santos 2008, 27–36). His privileging of form and totality over process offers something, too, to more recent discussions of dialectical method in geography, particularly attempts to reconcile Marxist and postcolonial approaches (Hart 2018). Santos, of course, as the itinerant trajectory of the Acknowledgments to this book attest, has a unique position in such intellectual histories, with his intimate personal position in relation to Third Worldism, anticolonial Marxism, geography, underdevelopment theory, and the postcolonial world. Much more research remains to be done to understand how Santos's work relates to and is in dialogue with—and perhaps challenges and reconfigures—the theoretical insights of Anglophone critical geography. It is interesting to note that Santos arrives at an insistence on the totality that somewhat recalls David Harvey's appeal to the "moral obligation" of geographers to overcome parochialism (Harvey 1974). For Santos, "geographers must lay the foundations of a truly human space" (Conclusion), and they must do so through a renewed focus on the totality. We might draw a connection here with the postwar renovation of humanism by anticolonial writers. Anticolonial thinkers such as Frantz Fanon, Aimé Césaire, and, more recently, Sylvia Wynter launched a fundamental critique of the false universalism of Western humanism, but, in so doing, they argued for a reconstitution of the category of the human, not its eradication. Santos's move is cognate: not the ceding of the totality but its reevaluation and reconstruction.

There are a few major features of Santos's work that this book does not introduce. First, though he touches on it, this book does not fully reflect his core interest in urbanization and urban geography, which can be read through books such as *Pobreza urbana* (1978), *A Urbanização Brasileira* (1993), and others. Second, Santos's specific training and practice in planning (Flavia Grimm in Bernardes et al. 2017) is only implicit here. Finally, Santos restrains from specific political or conjunctural interventions, remaining at the level of theory, epistemology, and ontology. His more directly political writings have, in fact, long been a neglected corner of his output, which has been rediscovered through two recent collections, *O país distorcido: o Brasil, a globalização e a cidadania* (2002), a set of essays published in the *Folha de S. Paulo,* one of Brazil's leading newspapers, in the 1980s and 1990s, and a collection of his writings for *A Tarde* between 1950 and 1960 (Santos 2019). These show a side to Santos that is quite distinct from the highly theoretical formulations that characterize *For a New Geography.*

There are many other avenues of thought that Santos's work opens up for Anglophone geography, many of which this book gives only a partial account of. We might cite, for instance, his discussion of Sartre's concept of the practico-inert and its possible application to analysis of infrastructural space and fixed capital (chapter 13) (see also Santos 2008, 80; 2009, 140), his insistence on the importance of periodization (chapter 18) (see also Santos 2009, 171–87), or his geographical account of the role of natural resources in the geopolitics of newly independent states (chapter 15). These are some of the tempting and intriguing elements of the book, ones that raise questions of missed opportunities in the history of Anglophone geography. They encourage impossible counterfactuals about how the time lag in the translation of Santos's work might have influenced the history of geography across languages. Though some of his ideas were published in English, we cannot know if geography might have developed differently had different geopolitics of knowledge and translation pertained and had Santos become as central to other-language geographies as he has been to Brazilian geography. There are, of course, serious shortcomings in the book that do not survive the test of time. The most striking of these, perhaps, is the committed methodological nationalism that Santos displays, prominently in chapter 16 but elsewhere too (though note that he changed his position on this later; see Santos 2008, 167–68). He sometimes takes the nation-state and its scale as a curiously fixed point of reference. Another strange feature of the book is the relatively limited engagement with histories of Brazilian thought, such as the geography of Josué de Castro, which Santos referred to elsewhere as foundational for his own work (Santos 2016) but does not reference here. Other readers will doubtlessly find other areas of disagreement, flaws, and oversights in Santos's writing.

Translation: Instance and Technique

As outlined earlier, the history of their (non-)translation is central to the reception—and nonreception—of Milton Santos's ideas within the history of Anglophone geography. The meanderings of his work across languages did not go only one way—we can identify key works that appeared first in French or English (Vasconcelos 2018, 53; and see for instance the essays collected in Portuguese in Santos 1979). However, here I want to turn to the complex process of translation itself. It is worth dwelling, here, on some elements of detail. Translating Milton Santos raises familiar challenges for translating from Portuguese into English, most notably that of subdividing long sentences. This raises some difficulties over maintaining the intricately connected elements of the units of Santos's thought, but in general his writing is clear, so these are manageable. There are, though, some substantive challenges of translation that this book elicits. These are important not least because they are central to the theoretical construction that is this work's driving concern. Though not in any way reducing problematics of translation merely to the lexical level, a number of words could be highlighted here as challenging for the translator—*rugosidade* as "roughness," *homem* as "man" or "people," among others—but I want to briefly discuss two whose translation into English crystallizes some theoretical concerns. The first is *instância*; the second is *técnica*.

Santos uses the term *instância* throughout the book, in more or less difficult ways. *Instância*, in Portuguese, has many of the meanings of "instance," in English, and can at times be translated by that English word. However, Santos's use of *instância* is not entirely reducible to this, as it retains a stronger sense than the English of process and ordering ("Instance, n." n.d.), drawing in particular on Althusser's writings and the debates around them in the 1970s (Contel 2014, 957). It is interesting to note that a similar issue arises with the translation of Manuel Castell's *The Urban Question,* by Alan Sheridan. While laying out some of Castell's own interpretation of a theory of space, Sheridan alternates between the terms "elements" and "instances" for categories of Castell's analysis. This leads to somewhat awkward results:

> The social organization of space may be understood, therefore, on the basis of the determination of spatial forms: By each of the elements of the three instances (economic, politico-juridical, ideological). These elements are always combined with the other element of their own instance. By the combination of the three instances. (127)

Sheridan's difficulties are my own, not least because they relate to two theorists of space and the urban, both of whom are making use of and intervening in debates within Marxist and Althusserian theory in the radical geography of the 1970s (for instance Peet 1978 shows others exploring similar issues at a similar time). The relation between "element" and "instance" is of some importance because it is a question of the relationship between structure, space, time and society. Santos's use goes beyond being a simply Althusserian reference point, and a translation needs to reflect that. The Hegelian and Marxian term "moment," which Harvey deploys significantly in *Limits to Capital,* displays some of the stakes here and is closely related. When Harvey characterizes the six "moments" of the socionatural totality, drawing on a footnote in *Capital* (Harvey 2008), these are conceived of as existing across space and time and must be dialectically related to avoid various forms of determinism. As Alex Loftus argues, this is a dynamic and dialectical conception that leads Harvey to a "method of moments" (Loftus 2012a, 17). In a more limited way, Santos's *instância* does some of the same work in his own context, and some have posited *instância* as one of *For a New Geography*'s key conceptual notions (de Toledo Junior 1996).

In light of this, it is significant that the Portuguese term retains more of the English term's older meanings of "insistence" and "command." Perhaps the most prominent example of this is in chapter 13 of the book, whose title in Portuguese is *Espaço como instância social*. This chapter revolves around a discussion of structural accounts of society, particularly drawing on Georgi Plekhanov, and argues that they have neglected space within their accounts of the functioning of society. Here the word *instância* is used to define what we might call forms of social structure or ways of ordering society. For example, Santos, in citing Franz Jakubowski developing Plekhanov, translates as *instância* what has been translated elsewhere as "levels" or "stages" of society (Jakubowski 1936).

These layers of translation are themselves worth attending to. A great deal of Santos's intellectual work is done through citation, and he is therefore a prolific translator within his own work. As readers will observe, Santos quotes extensively from sources in a number of languages and is generally conducting his own translations of these sources. There is conceptual work being done within these translations, as well as, in sequence, by my own. One example of the relevance of Santos's practice of translation illustrates the field here. In chapter 13 he translates a quotation from Harvey's *Social Justice and the City*. In the original (quoted in the English here, which is the edition Santos references), Harvey asks: "Should we regard urbanism as a structure which can be derived from the economic basis of society (or from superstructural elements) by way of a transformation? Or should we regard urbanism as a separate structure in interaction with other structures?" Setting aside, briefly, the broader questions of the conception of structure and superstructure, Santos translates Harvey's "urbanism" as "*a cidade,*" "the city." An alternative Portuguese word is available—*urbanismo*—but Santos chooses to denote "urbanism" as "the city." He also translates "structure" as *estrutura social,* which is somewhere between a gloss and a tweak. Elsewhere he translates Harvey's phrase "built environment" as *espaço criado* (see a footnote in chapter 12, "The Role of Roughness" section). Just as "built environment" is an important delineation in Harvey's work, so *espaço criado*—literally, created space— is in Santos's. It begins, for example, a movement toward his later concept of "used territory." But "created space" and "built environment" are not cognate. Santos is moving Harvey's work on urbanization into a theoretical vocabulary about space as such. Across languages there is a shift from a more descriptive to a more analytical category. With the backdrop of their two projects in mind, the traffic of ideas across languages enables what we might call slippages, or what we might call accretions. Indeed, we can note Santos's own apparent hesitation at this slippage, as within the very same footnote he translates Harvey's "built environment" differently, this time as the slightly more literal "*meio ambiente artificial*": "artificial environment." These details raise questions about how he is reading Harvey and how he conceptualizes the city as process or structure, topics that were at the forefront of his mind while writing this book, as he was concurrently working on his major work of urban geography, translated into English as *Shared Space: The Two Circuits of the Urban Economy in Underdeveloped Countries* (published in 1975 in English and recently reissued by Routledge). When translating Santos—and of course others—it is important to remember that translation was a central, if unacknowledged, part of his own thought processes. But his translations of Harvey also return us to the question of *instância*: there is an analytical distinction to be made, in English, among "level," "instance," and "order," so a choice has to be made at a lexical level.

Underpinning this discussion is a metatheoretical decision as a translator: in such translation, should my concern as a translator be for the source of Santos's concepts—Harvey, Jakubowski, or Plekhanov—or for Santos's use of them? The two are self-evidently interlinked, but as a task of translation it is at times necessary to privilege one or other in determining the English. Santos's and my own Russian is inadequate to discuss what Jakubowski originally wrote, but Santos translates him from the French, translated by Jean-Marie Brohm (Jakubowski 1976). There, the word is the French *niveau*, which quite easily translates into English as "level." This presupposes a more schematic, hierarchical conception than the *instância* that Santos deploys. Santos uses a cognate Portuguese term, *nível*, to describe Plekhanov's own schema, but *instância* for Jakubowski. The shift is necessary, intellectually, to distinguish between a hierarchical and static system and what Jakubowski calls a 'monist' (Jakubowski 1976, 96) conception of interrelated elements of society. This latter word, "element"—which I use to translate *instância* in this context—also gestures toward Santos's later conception of spaces as chemical processes that appears in chapter 18. Element, therefore, would seem to be a useful, "relevant" (Venuti 2003) translation of *instância*, and I deploy it there. In translating the title of this chapter, however, I have chosen to translate *instância* as "order," because I understand Santos to be drawing on the kind of structuring force and necessity that the English word "order" references. Translating one Portuguese word differently in different places is not an ideal solution, but translation is always a process of negotiation, and it is often messy.

This messiness, and this multiplicity of lexical transcription, brings me to the second example through which I want to draw out the theoretical stakes of translating Santos's work: the concept of *técnica*. This word, too, I have felt necessary to translate differently at different moments.[1] The intersecting notions of the technical/technics/technique/technology are vital to Santos's theoretical project. In his work they emerge from the history of French regional geography into a broadly Marxist theoretical framework for interpreting the production of space. The beginning of the first part of *A natureza do espaço* shows both the importance of *técnica* to Santos and the difficulty of translating the term into English:

> It is well established that the principal form of the relationship between man and nature, or better, between man and the environment, is through *técnica*. *As técnicas* are a set of instrumental and social means by which man realizes his life, produces, and, at the same time, creates space. This way of seeing *a técnica*, however, has still not been fully explored. (Santos 2009, 29)

Técnica, with or without the definite article, and its plural, *as técnicas*, is a core topic of that later book. It is therefore worth citing *A natureza do espaço* here to show that *For a New Geography* does not exhaust his inquiry into this category but lays the groundwork for his own and others' continued work on it as a central object for geographical enquiry. Although Lucas Melgaço and Carolyn Prouse (and others) have translated the terms with the English "technique/s" (Santos 2017), I would argue that his notion includes not only technology and (sometimes) techniques but also a broader philosophical concept that pertains to ways of doing that intercede between practice and theory. That is to say: technics. In this book he begins to outline the significance of technics and how the concept emerges from an older tradition of regional geography and *genre de vie*.

This is a significant set of theoretical connections not least because they emerge from a deep reading of the history of geography and Marxist theory. As early as chapter 1 of this book, in Santos's hands the analytical tools of *genre de vie* are tied to those of the mode of production to start a movement toward a critical geography. The question continues to be important. In chapter 15 he cites Bukharin when arguing that *técnica* is a key starting point for social analysis. The 1925 English translation of Bukharin has the quotation as follows: "technology is a varying quantity, and precisely its variations produce the changes in the relations between society and nature; technology therefore must constitute a point of departure in an analysis of social changes" (Bukharin 2017). Yet even Bukharin's argument revolves around key distinctions between a broadly conceived technics and a narrower definition of technology. Bukharin provides a footnote arguing that previous authors have "confuse[d] the method of production with technology" (Bukharin 2017). Here too a broader concept of technics is necessary.

In *A natureza do espaço* technics is a central concern, and Santos's reading of Marx deepened and moved across his life. Santos draws on French geography, in which he highlights that a few geographers—Pierre George, Pierre Gourou, and Maximilien Sorre—have engaged with the problem of technics in various important ways (Santos 2009, 30–35). These geographers, particularly the last, are all important inspirations and interlocutors for Santos in *For a New Geography*. One of the key features of his treatment of technics is its interconnection with periodization. This is one of Santos's enduring interests; he argues for the need to identify spatiotemporal epochs associated with shifting modes of production as a set of starting points for analyses of space itself. The definition of these epochs, Santos writes, emerges from analysis of technics: "Technology becomes history through the intermediation of technics. Technics are the modifying intermediation between nature and human groups" (chapter 9). This is an extension of an idea from volume I of Marx's *Capital*, where we find the argument that different economic periods are defined not by what is produced but by how it is produced (Santos 2009, 56).

In an interview from 1993 Santos stated the case clearly for this continuing methodological preoccupation:

The question of time and the materiality of space must be studied through the technical problematic. It is technics that give definition to materiality. We could even deploy a bit of poetic licence to say that nature itself can be studied from the perspective of the technical. (Santos 2008, 170)

We can see here that the word "technical" is tending toward the more abstract "technics" than the more limited "techniques" or "technology" by virtue of the breadth and scope of what is at stake. This becomes clearer when Santos continues:

We have, therefore, the time of actions, and, on the other hand, the time of materiality. It is thus that I conceive of the association of the ideas of time and space. Actions are an empty or concrete possibility available at a precise moment in history—my actions today are not the same as the actions I could have realized twenty years ago; actions are dated. But it is opportunity that makes possibility concrete, and opportunity is configured by materiality, which is ever more technical. This is the key to an epistemology of contemporary geography. (Santos 2008, 170)

This passage gestures toward the trajectory of Santos's thought after *For a New Geography.*

Technics—and technology—have been the subject of interest in radical geographical thought (Kirsch 1995), but much less than might be expected and in a less conceptually directed manner than Santos's work enables. Certainly, little of the geographical work that has considered the relation between theory and praxis (Staeheli and Lawson 1995; Minca 2000; Loftus 2012b, 2015) has taken up technics as relevant to this question. A theoretical concern with technics has a number of contact points with contemporary geographical enquiry, for instance the broad and growing field of infrastructural studies. The way in which infrastructures produce spaces and times has been the subject of a great deal of research, and Santos's particular constellation of references and ideas can enrich and stretch this field and its relationship with long-standing geographical concern to analyze the production of space. There is not room to further explore these connections here, but my central point here is that these theoretical questions are also questions of translation. Santos himself insists on the distinction between "particular techniques examined in all their singularity, and the technical phenomenon seen as a totality" (Santos 2009, 36). It is to draw out this distinction and to bring to bear the weight of philosophical and methodological reflection that comes with Santos's use of the term *técnica* that I therefore generally translate it using the slightly arcane but, I hope, precise English term "technics."

Conclusion

The preceding discussion of technics captures both the value and the difficulty of translating Santos. It opens out on the theoretical contribution that his work can make to Anglophone geography. Santos cuts a distinctive trajectory through critical geographical thought. He provides new and other ways of doing radical geographies and investigating the production of space and nature. As I hope this introduction has made clear, this book is not the final word on a theoretical system but the opening salvo of a distinctive, creative, and richly conceptual body of geographical thought. In spite of the solidity of his achievement, the recent Anglophone return to Milton Santos's work should not aim to simply put another brick in the edifice of the geographical canon. It should, rather, be a reminder of the underappreciated breadth of the history of thinking geographically about space, nature, and society beyond the English language and beyond the hegemonic whiteness and masculinity of our intellectual histories. It should be an invitation to keep looking beyond the enclosures of the canon to a much wider landscape.

Final Comments

At times, Santos's inconsistent referencing has required me to make a judgment as to the source he is referring to. There are some references missing from his original bibliography, and where they can be identified they have been included there. Where it has been impossible to identify a source or where a lack of page referencing makes it impossible to find a published English translation of a reference, this has been made clear in the notes. For example, the references to Manuel Castells are irregular, and it is not clear whether they are coming from the English edition or the French edition, as both are cited and both dates given. I have therefore homogenized these references to the English edition. Where Santos's bibliography appears to contain small errors, either over dates or titles, these have been corrected here. His mode of referencing—which includes both an author-date system and more extensive footnotes—has, however, has been retained. Where published English translations of sources originally quoted by Santos from other languages have been used, these have been added to the bibliography.

Bibliography

Barnes, Trevor J. 2017. "Retheorizing Economic Geography: From the Quantitative Revolution to the 'Cultural Turn.'" In *Theory and Methods*. Ed. C. Philo, 53–72. New York: Routledge.

Bernardes, Adriana, Adriano Zerbini, Cilene Gomes, Edison Bicudo, Eliza Almeida, Flavia Betioli Contel, Flávia Grimm, et al. 2017. "The Active Role of Geography: A Manifesto." *Antipode* 49, no. 4: 952–58. https://doi.org/10.1111/anti.12318.

Bukharin, Nikolai. 2017. *Historical Materialism*. Redditch: Read Books Ltd.

Castro, Ines E. de. 2002. "A região como problema para Milton Santos [The Region as a Problem for Milton Santos]." *Scripta Nova. Revista Electrónica de Geografía y Ciencias Sociales* 6, no. 124. http://www.ub.edu/geocrit/sn/sn-124e.htm.

Castro, Josué de. 1946. *Geografia da fome: A fome no Brasil* [Geography of Hunger: Hunger in Brazil]. Rio de Janeiro: Emprêsa Gráfica "O Cruzeiro" S.A.

Castro, Josué de. 1966. *Death in the Northeast: Poverty and Revolution in the Northeast of Brazil*. New York: Random House.

Contel, Fabio Betioli. 2014. "Milton Santos." In *Intérpretes do Brasil: clássicos, rebeldes e renegados*. Ed. Lincoln Secco and Luiz Bernardo Pericas. São Paulo: Boitempo Editorial.

Craggs, Ruth, and Hannah Neate. 2019. "What Happens If We Start from Nigeria? Diversifying Histories of Geography." *Annals of the American Association of Geographers* 110, no. 3: 1–18. https://doi.org/10.1080/24694452.2019.1631748.

Daigle, Michelle, and Margaret Marietta Ramírez. 2019. "Decolonial Geographies." In *Keywords in Radical Geography: Antipode at 50*. Ed. Antipode Editorial Collective, 78–84. John Wiley & Sons, Ltd. https://doi.org/10.1002/9781119558071.ch14.

Davies, Archie. 2018. "Milton Santos: The Conceptual Geographer and the Philosophy of Technics." *Progress in Human Geography* (January): 1–8. https://doi.org/10.1177/0309132517753809.

Doherty, Joe. 2020. Interview by Archie Davies.

Esson, James, Patricia Noxolo, Richard Baxter, Patricia Daley, and Margaret Byron. 2017. "The 2017 RGS–IBG Chair's Theme: Decolonising Geographical Knowledges, or Reproducing Coloniality?" *Area* 49, no. 3: 384–88.

Featherstone, David. 2019. "New Left." In *Keywords in Radical Geography: Antipode at 50*. Ed. Antipode Editorial Collective, 192–97. John Wiley & Sons, Ltd. https://doi.org/10.1002/9781119558071.ch35.

Ferretti, Federico. 2019. "History and Philosophy of Geography I: Decolonising the Discipline, Diversifying Archives and Historicising Radicalism." *Progress in Human Geography*, December, 0309132519893442. https://doi.org/10.1177/0309132519893442.

Ferretti, Federico, and Breno Viotto Pedrosa. 2018. "Inventing Critical Development: A Brazilian Geographer and His Northern Networks." *Transactions of the Institute of British Geographers* (May). https://doi.org/10.1111/tran.12241.

Gappert, Gary, and Harold M Rose. 1975. *The Social Economy of Cities*. Beverly Hills, Calif.: Sage.

Grimm, Flavia Christina Andrade. 2011. "Trajetória epistemológica de Milton Santos. Uma Leitura a partir da centralidade da técnica, dos diálogos com a economia política e da cidadania como práxis [The Epistemological Trajectory of Milton Santos: A Reading from the Centrality of Technics, of Dialogues with Political Economy and Citizenship as Praxis]." Ph.D. diss., Universidade de São Paulo.

Hart, Gillian. 2018. "Relational Comparison Revisited: Marxist Postcolonial Geographies in Practice." *Progress in Human Geography* 42, no. 3: 371–94. https://doi.org/10.1177/0309132516681388.

Harvey, David. 1974. "What Kind of Geography for What Kind of Public Policy?" *Transactions of the Institute of British Geographers*, no. 63: 18–24. https://doi.org/10.2307/621527.

Harvey, David. 2008. "On the Deep Relevance of a Certain Footnote in Marx's Capital." *Human Geography* 1, no. 2. https://hugeog.com/onadeeprel/.

"Instance, n." n.d. In *OED Online*. Oxford University Press. Accessed 7 January 2020. https://www.oed.com/view/Entry/97031.

Jakubowski, Franz. 1936. *Ideology and Superstructure in Historical Materialism*. London: Pluto Press.

Jakubowski, Franz. 1976. *Les superstructures idéologiques dans la conception matérialiste de l'histoire*. Paris: Editions de l'Atelier.

Jazeel, Tariq. 2014. "Subaltern Geographies: Geographical Knowledge and Postcolonial Strategy." *Singapore Journal of Tropical Geography* 35, no. 1: 88–103.

Kirkendall, Andrew J. 2004. "Entering History: Paulo Freire and the Politics of the Brazilian Northeast, 1958–1964." *Luso-Brazilian Review* 41, no. 1: 168–89. https://doi.org/10.3368/lbr.41.1.168.

Kirsch, Scott. 1995. "The Incredible Shrinking World? Technology and the Production of Space." *Environment and Planning D: Society and Space* 13, no. 5: 529–55.

Loftus, Alex. 2012a. *Everyday Environmentalism: Creating an Urban Political Ecology*. Minneapolis: University of Minnesota Press.

Loftus, Alex. 2012b. "Gramsci, Nature, and the Philosophy of Praxis." In *Gramsci: Space, Nature, Politics.* Ed. Alex Loftus, Michael Ekers, Gillian Hart, and Stefan Kipfer, 178–96. Chichester: Wiley-Blackwell. https://doi.org/10.1002/9781118295588.ch9.

Loftus, Alex. 2015. "Political Ecology as Praxis." In *Routledge Handbook of Political Ecology.* Ed. Tom Perreault, Gavin Bridge, and James McCarthy, 179–87. London: Routledge.

Maddrell, Avril. 2008. "The 'Map Girls.' British Women Geographers' War Work, Shifting Gender Boundaries and Reflections on the History of Geography." *Transactions of the Institute of British Geographers* 33, no. 1: 127–48.

McKittrick, Katherine. 2006. *Demonic Grounds: Black Women and the Cartographies of Struggle.* Minneapolis: University of Minnesota Press.

Melgaço, Lucas, and Carolyn Prouse. 2017. *Milton Santos: A Pioneer in Critical Geography from the Global South.* Vol. 11. New York: Springer.

Minca, Claudio. 2000. "Venetian Geographical Praxis." *Environment and Planning D: Society and Space* 18, no. 3: 285–89.

Moreira, Bruno de Oliveira. 2010. "'Visita a uma revolução': Uma análise dos escritos de Milton Santos sobre a revolução cubana (1960)." *Revista de História (UFBA)* 2, no. 1. https://portalseer.ufba.br/index.php/rhufba/article/view/27693.

Pedrosa, Breno Viotto. 2015a. "A controvérsia da geografia crítica no Brasil (parte I) [The Controversy of Critical Geography in Brazil (Part 1)]." *Geosul* 30, no. 5: 7–28.

Pedrosa, Breno Viotto. 2015b. "A controvérsia da geografia crítica no Brasil (parte II) [The Controversy of Critical Geography in Brazil (Part 2)]." *Geosul* 30, no. 5: 29–46.

Peet, Richard. 1978. "Materialism, Social Formation and Socio-Spatial Relations: An Essay in Marxist Geography." *Cahiers de géographie du Québec* 22, no. 5: 147–57. https://doi.org/10.7202/021390ar.

Peet, Richard. 1985. "Radical Geography in the United States: A Personal History." *Antipode* 17, no. 2–3: 1–7. https://doi.org/10.1111/j.1467-8330.1985.tb00323.x.

Power, Marcus, and James D. Sidaway. 2004. 'The Degeneration of Tropical Geography'. *Annals of the Association of American Geographers* 94, no. 3: 585–601. https://doi.org/10.1111/j.1467-8306.2004.00415.x.

Santos, Milton. 1975. 'Underdevelopment, Growth Poles and Social Justice'. *Civilisations* 25, no. 1/2: 18–32.

Santos, Milton. 1979. *Economia espacial: Críticas e alternativas.* Ed. Maria Irene de Q. F Szmrecsanyi. São Paulo: EDUSP.

Santos, Milton. 1980. "The Devil's Totality: How Geographic Forms Diffuse Capital and Change Social Structures." *Antipode* 12, no. 3: 41–46.

Santos, Milton. [1994] 2008. *Técnica, espaço, e tempo: Globalização e meio técnico-científico-informacional.* São Paulo: EDUSP.

Santos, Milton. [1996] 2009. *A natureza do espaço: Técnica e tempo, razão e emoção* [The Nature of Space: Technics and Time, Reason and Emotion]. 4th ed. São Paulo: Universidade de São Paulo.

Santos, Milton. 2016. "Entrevista explosiva com Milton Santos [Explosive Interview with Milton Santos]." *Geledés* (blog). 28 December. https://www.geledes .org.br/entrevista-explosiva-com-milton-santos/.

Santos, Milton. 2017. *Toward an Other Globalization: From the Single Thought to Universal Conscience.* Vol. 12. New York: Springer.

Santos, Milton. 2019. *Milton Santos: Correspondente do jornal A Tarde.* Ed. Maria Auxiliadora da Silva Silva and Willian Antunes. Florianópolis: Série Livros Geográficos.

Santos, Milton, and Adriana Bernardes. 1999. "Tarefas da geografia Brasileira num mundo em transformação: um momento de sua trajetória." *Revista Ciência Geográfica, AGB/Bauru* 13: 4–22.

Sharp, Joanne. 2019. "Practising Subalternity: Postcolonial Tanzania, the Dar School and Pan-African Geopolitical Imaginations." In *Subaltern Geographies.* Ed. Tariq Jazeel and Stephen Legg. Athens: University of Georgia Press.

Staeheli, Lynn A., and Victoria A. Lawson. 1995. "Feminism, Praxis, and Human Geography." *Geographical Analysis* 27, no. 4: 321–38.

Theodore, Nik, Tariq Jazeel, Andy Kent, and Katherine McKittrick. 2019. "Keywords in Radical Geography: An Introduction." In *Keywords in Radical Geography: Antipode at 50.* Ed. Antipode Editorial Collective, 1–13. Chichester: John Wiley & Sons, Ltd. https://doi.org/10.1002/9781119558071.ch1.

Toledo Junior, Rubens de. 1996. "O espaço como instancia social: A base para uma geografia nova." In *Ensaios de Geografia Contemporanea: Milton Santos Obra Revisitada.* Ed. Ana Fani Alessandri Carlos. São Paulo: Hucitec.

Vasconcelos, Pedro de Almeida. 2018. "Milton Almeida Dos Santos (1926–2001)." In *Geographers: Biobibliographical Studies.* Ed. Elizabeth Baigent and André Reyes Novaes, 41–68. London: Bloomsbury.

Venuti, Lawrence. 2003. "Translating Derrida on Translation: Relevance and Disciplinary Resistance." *Yale Journal of Criticism* 16, no. 2: 237–62.

FOR A NEW GEOGRAPHY

Introduction

From a Critique of Geography to a Critical Geography

In 1925, in the preface of his famous treatise, Emmanuel de Martonne ([1925] 1957, vol. 1, 20) wrote that geography could now be considered a fully fledged science. Forty years later, as new paradigms came to dominate the discipline, Peter Haggett and Richard J. Chorley (1965, 371) were no less categorical. In terms of geography's scientific qualities, they argued, "one initial problem is quickly resolved. To ask whether geography is or is not a science is like asking whether sports are games." I follow Brunhes (1910) on a different tack: "human geography has not yet been made, we still have to make it."

A New Geography?

Putting forward a *New Geography* might, at first, seem hugely pretentious, as if I were claiming to have invented something. But everything, including the sciences, is subject to constant processes of regeneration. The new is not invented; it is discovered.[1]

When the general conditions of life on earth change, or when interpretations of humanity and matter change, all scientific disciplines have to reconfigure themselves. They need new languages to express their part of total reality.

We are living through one of those periods when the meaning of things undergoes a revolutionary transformation. Some disciplines perceive these qualitative changes fully and incorporate them into their body of knowledge. Others do so partially and incompletely. This undermines systematic

1

inquiry: you cannot structure a coherent analysis when some things are analyzed in light of a new paradigm and others with theories that are out of date. This is where Geography is today.

Indeed, it was ever thus. Ever since the end of the nineteenth century and the foundation of what has historically been called "scientific geography," it has been impossible to construct a set of propositions based on a common system and connected by an internal logic. This is because geography has always been more concerned with a narcissistic debate over geography as a discipline than with dealing with geography as an object. Up to today, *geography* has been discussed more than *space,* the object of geographical science. Geography's conceptual approach has been built from outside the object of the science, rather than from within it.

This has led to a grave epistemological error. Attempts to make conceptual progress have imported insights from affiliated disciplines, rather than starting from those aspects of reality that geography examines. The consequent buildup of errors has complicated the task of setting an intellectual trajectory that can attribute to geographical space—the object of geography—a set of basic principles that can guide theory, empirical research, and action.

But this task can now begin. Philosophy has abandoned its role as the gatekeeper of science and come to concern itself with ideas and their coherence. We need no longer refer upward to a generalized philosophy that dictates norms of thought and ascribes a teleology to each particular discipline. Each discipline has acquired its own epistemology—what Bachelard has called "regional theory"—founded in its own practice and referring to its own object. Of course, we are not looking to establish an independent science, because no science is truly independent. Social reality is singular, and it falls to each science to study one of its aspects. But this does not invalidate the notion of the unity of science, since to study totality through totality can only lead to tautology.

As any form of productive activity becomes more complex, a division of labor emerges. Hence the development of distinct and autonomous sciences. Each takes as its object a part of social reality. In a continuously developing process operating at multiple levels, each science adopts its own general principles and norms of proceeding, its own epistemologies and technologies. But autonomy is not independence. Each science creates its own particular universe: a system for thinking through one part, one aspect of the thing. But these particular universes must be subordinated to a holistic approach appropriate to reality as a totality. Each particular science is not merely the result of an arbitrary splitting up of a science of the total thing—some theoretical "total science." Equally, the object that each science attributes to itself cannot at any moment be arbitrarily segmented.

An Ambitious Project

This book aims to be the first of a series of five under the general theme of *Human Space*. In spite of huge amounts of work by different specialists—above all, geographers—for more than a century, there has, with the exception of the magisterial work of Maximilien Sorre, been little attempt to outline a totalized system of human space. The task I have set myself, under a unique set of personal and historical conditions, is therefore both arduous and audacious. Its inherent difficulty leads me to declare early on that I have spent a lot of energy on this in the past few years, with undoubtedly modest results. But this is no reason to exempt myself from the responsibility of sharing my teaching and research. It emerges from a period of working in, and confronting, the diverse realities of countries and cultures in North, West, and East Africa; Europe; and the Americas. As an interdisciplinary project, it has obliged me to read beyond the bounds of geography, in both traditional and modern social sciences. It has obliged me to take an interest in philosophical categories and principles of the exact sciences that I previously could not have imagined would be useful for establishing better knowledge of human space. My greatest challenge, however, has been to find a form of expression that is at once exact and simple. The reader will judge whether I achieved this objective.

The set of works of which this is the first must form a coherent whole. However, each volume should be independently readable as a book. This has obliged me to set out an unavoidably complex schema. The process of writing brings new ideas and new knowledge to light. It is likely, therefore, that the plan for subsequent volumes will change, just as this book has itself transformed from its original drafts.

Searching for a critical geography, this book first of all critically analyzes the evolution of geography. Many other authors have investigated this subject, and I do not pretend to have exhausted it. My objective is merely to point toward those problems that, in my view, stand in the way of the construction of a geography attuned to a more open and constructive social problematic. This critique is neither partial nor gratuitous but has the specific aim of introducing my small contribution to building that long sought-after critical geography. I set out from the past with a view to the future.

The four later volumes will be dedicated, respectively, to the following themes: (2) *From Cosmic Nature to the International Division of Labor*; (3) *Spatial Organization and Contemporary Society*; (4) *Social Time and Human Space*; (5) *Social Totality and Total Space: Form, Function, Process and Structure*. This list does not mean these books will appear in this order.

The second volume, provisionally entitled *From Cosmic Nature to the International Division of Labor*, will explain what we could call the process of the production of space. Its thesis is that, in becoming a producer—that is, in the conscious utilization of the instruments of labor—man becomes at once a social being and a creator of space. Spatial evolution occurs thanks to the coming together of factors of production and relations of production, marked, through time, by the diverse stages of the international and internal division of labor. The extension of the division of labor corresponds to the separation, in space, of different parts of the productive process, which themselves take on different values in different eras. Urbanization, like other forms of spatial distribution, is a result of one stage of this process. The study of the production of space should operate as a theory of human space.

The third volume will deal specifically with the *Spatial Organization of Contemporary Society*. It will discuss the economic, social, and political forms of the contemporary moment: the *spatial present*, taken as a specific historical reality. An attempt to define the technological era and the universalization of society that it engenders will lead to a definition of its consequence: *global, total space*. Analytically, and geographically, the unit of study will be the nation-state. Among other themes, the book will reinterpret the phenomenon of urbanization, with special reference to underdeveloped countries.

The fourth volume will deal with the relations between (social) time and (total) space. The notion of social time raises the question of the periodization of history. This epistemological necessity emerges from the fact that history is at once continuous and discontinuous. The category of modes of production enables a periodization of history. It is, however, insufficient, because within time there are many temporalities. The time of the mode of production is universal; from it emerges the need to address the temporality of the nation-state. This argument underwrites an analysis of the articulation between the international and the internal divisions of labor and of the interplay between the internal and external forces of the formation of society and space. The notion of empirical time is the only thing that can be reconciled with the idea of objective space. It enables us to work in terms of space-time systems. This volume will be a first attempt at the construction of an epistemology of human space, emerging from the theory laid out in the first three volumes. The fifth volume will complete this epistemological undertaking.

That fifth and final volume of the series will deal with problems that can be considered, generically, to belong to a dialectic of space or, rather, a dialectic in space. The social totality is treated as a being whose existence is, in the end, given through total space. The study of the social totality in a permanent process of totalization is associated, therefore, with an analysis of space in a permanent process of change. Social change leads to ruptures that profoundly alter the organization of space. These two interconnected movements can be analyzed only through categories that are simultaneously categories of reality. Form, function, process, and structure are, therefore, treated as both analytical and real categories that are interdependent and mutually imbricated. The role of ideology will be defined both within the social totality and within space. Landscape is a kind of "functional lie." Only through studying the movement of the totality can we distill ideology. This lets us define structure, context, and tendency. Perhaps in this way we can reconstruct the future, in an epoch in which space has become a fundamental philosophical and political category.

If, as I hope, this becomes a coherent project, these disparate elements will eventually constitute a whole and the themes will interlace. Such repetitions as are necessary for clarity will not interfere with each book's own focus.

In this volume I will deal with themes including time, relations between form and function and between process and structure, and contemporary spatial organization. For the sake of clarity, at times I will touch on things that will be covered later in more breadth and depth. I will attempt to minimize repetition.

The preoccupation that guides this book—the first stage of the task I have set myself—is to take the problematic of space up by the root. I therefore begin by analyzing the work of different schools of geographical thought in order, ultimately, to propose a project of study based on contemporary realities. This will be at once a theory and an epistemology.

I aim to provide both an explanation of spatial reality and the instruments for its analysis. An epistemology that is not based on a theory is ill made, because it offers instruments of analysis that misconstrue or deform reality. Equally, a theory that does not generate its own epistemology is useless because it cannot be put to work. Scientific coherence cannot be achieved in any other way.

A Necessary Risk

I know this is a risky business. Bertrand Russell wrote that "any doctrine with some kind of coherence is, surely, at least in part, irksome and contrary to current prejudices" (1965, 93).[2] Science can be co-opted by vested interests. It often deliberately misleads in order to benefit them. If it is to face up to the current state of knowledge, scientific work must be political work.

To renew science is to renew social forms of thought. This is as true today as it was in the time of Galileo. Perhaps even more so. Any attempt to renew a science, to tune it into the real, will face many obstacles. These will be most difficult to overcome when they emerge from within the discipline itself. Scientific errors have a professional seal of approval that gets stronger the longer they persist. Correcting them risks wounding colleagues' pride. However, it is not enough merely to nod along with Robert Lynd that this is a critical time for the social sciences: this is no time for politesse.[3]

How the scientific collective behaves is key to the diffusion of ideas, particularly when they seem new or shocking. The most experienced struggle most to accept the new. As Haggett has noted (1965, 114): "students are often more ready to receive new ideas than we are ready to teach them." As Bernardo Secchi warns, "such an endeavour is unpopular if it has fatuous origins and the author presents it as a definitive interpretation of past endeavours and the only guide to future ones."[4] This should not slow us down, because science cannot move forward without healthy critique, and there is no critique without risk.

Basic categories like man, nature, and social relations will always exist as instruments of analysis, but they change with each historical period. The past cannot be the master of the present. All groundbreaking work requires an enormous effort from its author to shrug off the weight of inherited knowledge, because the new is still not made and still not codified. The new is the unknown. It can be conceptualized only with the imagination, and not through mere conviction.[5] We must be willing, therefore, to present what we might call our *proto-ideas* as outcomes of our work. These are vital if others are to participate in our inquiries. The transmitted idea is always a codification, an imprisonment of a concept by a language. The proto-idea is an idea in the midst of its making, one that inserts the reader into the process of its production.

We can use a Kantian notion: "when we compare an author's thoughts on a subject that they study, it is easy to think that we understand it better than they do themselves" (Kant 1929, 310). The making of the idea precedes the discovery of the language necessary to express it. The creator of an idea works with the vocabulary available to him: a lexicon designed to express one set of thoughts that he wants to substitute with another.

This can bring the thinker premature satisfaction. Rather, we should follow Frederick Woodbridge (1940, ix) in *An Essay on Nature*: "In what I have written I have been deeply serious, but often I have had to smile at my professions of authority."

PART I

The Critique of Geography

1

The Founders

Scientific Pretensions

Geography, wrote Jean Dresch in 1948, was "born not during the growth of the bourgeoisie but in the very flush of its triumph":

> In the beginning, it was as much a philosophy as a science. A philosophy that, like German historians, German geographers made to serve political ends. It was used as a medium for national and international propaganda, and as a weapon of war between States and Empires. Perhaps even more than History was. Geography still bears the marks of its youth and of the social, economic, and political conditions in which it developed. By virtue of its own methods, geography—more than any other science—bears the weight of current ideology. (88)

In fact, official geography was "from its beginnings" more an ideology than a philosophy, not only in Germany but elsewhere. Indeed, Dresch recognizes this when he writes that "from its origins [geography] was necessarily directed by ideology."

What, then, was this ideology?

The Ideology of Geography

From the start, capitalist ideology has had to be equal to capital's need for expansion into both centers and peripheries. In its early years of growth, capitalism needed to remedy excesses in both production and capital, as well as to restrain the social and economic crises that shook the countries of the center. It needed to create the conditions of possibility for the expansion of commerce. Industry's thirst for raw materials led to the opening of mines overseas and the conquest of lands for food production. In this period, the international division of labor gained a new dimension. The spatial and economic structures of poor countries were remade to ensure ceaseless devotion to their tasks within a new global division of labor. In this transformation, geography was called upon to play a crucial role.

Geographers were divided over the triumphant forward march of imperialism. On one side, there were those, such as Élisée Reclus and Camille Vallaux, who struggled for a more just world and for the organization of space to provide man with greater happiness and equality.[1] Peter Kropotkin was in this camp. He saw space as key to constructing a new society. (It is of no significance that the anarchist prince was not officially a geographer.)

On the other side, there were those who advocated for colonialism and on behalf of the empire of capital. There was also a larger group who thought of themselves as humanists. They, however, were incapable of constructing a geographical science that conformed to their generous desires.

Coming late into the world as an official science, geography has struggled since its infancy to disconnect itself from powerful interests. One of the great conceptual feats of geography was to hide the role of the state—and of class—in the organization of society and space. Its other was to justify the colonial project.

Colonial Geography

The use of geography as a tool of colonial conquest was not isolated to one country. In all colonizing countries, under various conditions, geographers were corralled to a task that gained new impetus every time history stirred. T. W. Freeman (1961, 9) argues that the expansion of geography and the expansion of colonization are connected.[2] The impetus our discipline gave to colonization and the role it played in it were factors in how each developed. The notable Nigerian geographer Akin Mabogunje insisted on this in his 1975 work, though he slightly overplayed his hand in terms of French geographers and underemphasized the English.

The first chair in geography in France was established in Paris in 1809. In 1889, it passed to Paul Vidal de La Blache after the death of Auguste

Himly. The second chair, created in Paris in 1892, was in colonial geography. It was occupied by Marcel Dubois. The next chair, specifically of colonial geography, was established in 1937 and was held initially by Charles Robequain. Other chairs in the same subject—beyond those initiated in 1899 at the National School of Overseas France—were founded in Bordeaux (1946), Aix-en-Provence, and Strasbourg.

Among the English geographers, Halford Mackinder was the most effective servant of imperialism. In France, even Paul Vidal de La Blache seemed enamored of the colonial project. In one of his articles, published in the *Annales de Géographie* [Annals of Geography], and afterward in his posthumous *Principes de géographie humaine* [Principles of Human Geography], he wrote that distant conquests had put man in a new situation:

> We must congratulate ourselves. The task of colonization—the glory of our age—would be trivial if nature were able to impose strict limits on us, rather than leave scope for that work of transformation and reconstruction whose realisation is within the power of man.

Albert Demangeon, in the conclusion of his classic 1947 work, wrote of the French colonial project in Sudan, "this Black Africa that still offers to European colonization a wonderful horizon" (395).

The Dutch, Belgians, and others would need to be part of any full accounting of imperialist geographers.

Determinism and Its Sequels

Determinism, region, *genre de vie,* and culture. These are all apparently disparate and innocent ideas, but they all follow a similar path.

According to Harry Elmer Barnes (1925, 49) the modern architect of determinism was the historian Henry Thomas Buckle, with his "model" to "make history an exact science."

Griffith Taylor (1946, 4), despite advocating philosophical and religious factors in geographical understanding, is an arch "determinist." He sometimes even refused to include any cultural factors at all in the concept of environment (1951, 9). According to Brian Berry's *Urban Geography* (1970), Taylor produced "perhaps the most extreme geographical study of the city from an environmentalist point of view." We owe to Taylor insights like "commercial relations, for climactic reasons, are established more in a North–South direction than an East–West direction"; this reminds me of Herodotus: "the sources of the Nile were not habitable because of excessive heat."

In this regard, Ellen Churchill Semple's contribution was significant. Though Randle (1966) considers her work to be "ingenuous determinism," Broek (1967, 27) differs, lamenting that "unfortunately Miss Semple insists on natural relations and almost forgets the lessons of Ratzel." Indeed, even contemporary discourse among educated people often reduces geography to the interpretation of natural conditions.

And let's not forget Ellsworth Huntington: "temperate climates are excellent for civilization" . . . "Excessive heat is debilitating" . . . "Excessive cold is stupefying."

According to Michael Chisholm (1966, 15–16), the influence of geography's failures to correctly interpret the role of nature helps explain other disciplines' loss of confidence in it.[3] Maximilien Sorre (1957, 155) has argued that sociologists have been particularly badly affected.[4] According to Ray Pahl (1965, 84), "tacit determinism" led many geographers "to search so readily for the (implicit) influence of the physical environment in patterns of settlement and economic functions of society." This has implications for the study of the urban environment.

Cultural Geography and *Genres de Vie*

In 1962, Carl Sauer said that the study of geography was limited by the multiplicity of ways it could be approached. Two fundamental and opposing tendencies appeared. On one hand, one group claimed that its greatest interest lay in man—in the relations between man and his environment, largely in the sense of human adaptation to the physical environment—and another group directed its attention to the elements of material culture that characterize an area. Sauer argues that we can, for convenience, call the first position "human geography" and the second "cultural geography."[5] He adds that "the terms are employed in this manner, but not exclusively."[6]

Patrick Bryan's notion of a *cultural area* (1933) starts from landscape and ends up at regional subdivision. For him, the earth is a set of specific forms of the use of territory—of cultural areas—that are the result of the labor of different societies and are based in their cultural diversity. For J. W. Watson (1951, 468), this represents progress for the idea of the natural region and a step in the direction of social geography. However, this optic risks seeing landscapes as frozen frames. While society always gives landscape new functions, meanings, and values, as a frame of action itself landscape remains largely immutable.

The great American geographer Carl Sauer lamented that geographers on either side of the North Atlantic knew little about one another's work. This is still true. Sauer himself looked back to Harlan Barrows and his presidential speech to the Association of American Geographers in 1923. There, Barrows announced geography to be the ecology of man, and Sauer reaffirmed the dominance of this approach in the Anglophone world.

Sauer recognized that different tendencies had responses on either side of the Atlantic divide. Commentaries on Friedrich Ratzel, for instance, show the breadth of possible interpretations. This brought Camille Vallaux to the attention of American cultural geographers. The *Annals of the Association of American Geographers,* first published in April 1911, was interested in human geography. Sauer recognized, too, that, while different tendencies had transatlantic counterparts, they also had weak imitations.

Indeed, the school of cultural areas ran parallel to urban ecology. Without going into details, this school is really just regional geography in American disguise. Sorre was right when he said ([1955] 1962, 44–45) that the form of geographical explanation is ecological, emerging from the relations between beings and their environment, and not only "reciprocal relations because we are presented with a mass of complex actions, reactions, and interactions."[7]

The concept of *genre de vie,* or mode of life, proposed by Vidal de La Blache (1911, 193–212, 289–304) is one of the key paradigms that has oriented modern human geography. Here, through the intermediation of a series of modes of doing—technics—embedded in a local culture, man enters into relation with nature. As an object of study, space comes to be the result of an interaction between a localized society and a given natural context: an argument designed to reinforce the idea that the region should be the unit of geographical study.

Sorre (1948) has very justifiably objected that the concept of *genres de vies*, while useful for nondeveloped societies, was no longer applicable in the modern world in which human groups act principally according to drivers that come from outside. Has anyone paid attention? Old geographical ideas die hard. They are frequently abandoned, only to reappear in another form. Ecological approaches, just as much as the regional school, are at risk of determinism.[8]

"Ecologists" on both sides of the North Atlantic have adopted Hettner's inclusion of geography within the chorographic disciplines. Crucially, Sorre adds the category of history and the notion of time.[9] For Sorre, "historical and ecological explanation are the two types of explanation invoked by all the sciences of living things.... Historical explanation completes ecological explanation and prevents its excesses" (Sorre 1962, 44, 46).

The foundations of the Sauer school are similar to the ideas of Vidal de La Blache and his students. Cultural geography is possibilist, and the notion of *genre de vie*—associated with the region, according to Vidal—is not so far from the idea of the cultural area. In this sense, among the French geographers, Pierre Gourou perhaps best syncretized the two approaches, and among the Anglo-Saxon geographers, Robert Dickinson.

These two, among others, have introduced the concept of "civilization" as a framework for man and his context.[10] This retains *genre de vie* and combines it with a techno-*cultural* focus that disregards the technical-*economic* context. By suggesting that technics is connected to culture and not to the mode of production, this position compromises ongoing debates about underdevelopment.

If we use the term "civilization," wrote Jean-Jacques Goblot (1967, 73), "it is not to make it an operational concept or an instrument of analysis, but merely to designate (not even to define) the concrete reality of 'local' historical development, whose specific determinations constitute the true object of analysis."

We might think that human ecology, successfully introduced by a group in Chicago (Park and Burgess 1921), would have come to the rescue of a discredited geography (Stoddart 1967, 521) with a new paradigm. In the first phase, ecology offered a more advanced conceptual framework but did not move away from its old orientation.[11] This new discipline treated nature and man as if they were opposite categories. "Nature" as part of the ecological system is "primary" nature and not socialized nature. It is nature without human history. Man acts on the environment as if he were separate from it, and not one of its elements. This conception accentuated the mistakes of regional geography and perpetuated a dualism that penetrated other disciplines. It meant that a philosopher like Sergio Bagú (1973, 114–15), so rigorous in his original historical analysis, could write that a given society is equal to the population, plus the national global system, plus natural resources.[12] An equally eminent economist could write that social structure is not homogenous, because it is made up of the following structures: (1) Geographic and physical; (2) Demographic; (3) Technical and economic; (4) Institutional, social, psychological, and mental; (5) Cultural (Baltra Cortes 1966, 42–50). Even a sociologist of the caliber of George Dalton (1971, 89) could define an economic organization as "a set of rules in force through which natural resources, human cooperation, and technology are brought together to provide material items and specialist services in sustained and repetitive fashion."

The Ruin of Classic Geography

The idea of the region must be at the center of a renewed debate.[13] Can we, even today, hold that human constructions, as they appear on the face of the earth, result from an interaction between "a" human group and "its" geographical environment?

Sorre has already responded to this question in his writing on "derived landscapes." The landscapes of underdeveloped countries derive from the necessities of the economies of industrial countries where decisions are ultimately made. The relations between a human group and its geographical foundations do not depend only on that human group.

These very relations, enacted through intermediaries whose quality and nature vary, are themselves one of the sources that reinforce unequal social structures.

In the face of the exigencies of social and economic life, social sections or classes, either created or strengthened by the relations between underdeveloped countries or regions and developed countries or regions, behave differently. These diverse behaviors have many geographical consequences even in a single space. I have sought to show this in reference to food (Santos 1967): "the principle characteristics of a holistic geography of diet are strange, and at first sight paradoxical, because they denote a general geography that does not pass through a regional geography of a classical form." The foundations of a general geography of alimentation would be the diverse elements that characterize and define regions but that are not themselves regions. It would pass directly from the analysis of reality in particular sectors and subsectors of society and economy to a general geography.

This is obvious in both rural and urban areas in underdeveloped countries. Many distinct forms of social reality are overlaid in the same space, thanks to the division of people into classes with marked differences of income, consumption, quality of life, and so on.

There are geographical spaces whose characteristics derive from the intimate interaction between *a* human group and *a* geographical base, but they are less and less common. They appear to be the result of a lack of social dynamism frequently referred to in today's terminology as "geographical dynamism." They demonstrate a lack of response to the conditions of the modern world and local failures to adapt to the influence of economic and social progress.

Progress in transport and communications and the "globalization" and expansion of the international economy explain the crisis of the classical notion of the *region*. If we still want to retain that denomination, we must find a new definition for the word.

In the contemporary conditions of the world economy, the *region* is no longer a living reality characterized by internal coherence. It is, as Kayser observed, principally defined from outside. Diverse criteria alter its limits. Under these conditions, the region in itself has ceased to exist.[14]

A general geography based on a so-called region relies on false relations stripped of autonomy and explanatory capacity. It overemphasizes relations that merely intercede between human groups and the geographical environments in which they live. Any search for causality on these terms will lead inevitably to serious errors. It can lead to empirical abstraction, which valorizes "things in themselves" and not the relations that they embody or obscure. Diverse forms of mediation, among which we could count political, financial, commercial, or economic techniques, lend to human–environment relations another dimension that excludes the rigidity of a classical form of regional geography and the mechanicism of their relations with "general" geography. A valid theory cannot be established on the "principle of causality." The fact that there is no such thing as *regional autonomy* means the liquidation of traditional regional geography.

The Dangers of Analogy

The weakness of human geography, wrote Jean Gottmann,

> comes from a tendency to drink from the same spring as physical geography; natural history. We cannot expect similar behaviour from human collectives as from more elementary living beings. The simplistic determinism of botany can only allow us to scratch at the surface of the problems of human societies. (1947, 5)

Geographers have often worked through analogy, above all to the natural sciences. Two serious errors emerge from this. First, we cannot mechanically transpose what happens in the physical world onto what happens in history. Second, analogy leads us to examine objects from the outside, enabling only an understanding of their facets or form, when it is the *content* of objects that allows us to identify, individualize, and define them.

Historical phenomena never repeat themselves in the same form. Nor are the relations between different social groups the same in different periods. The laws of development, therefore, as Serafin Meliujin (1963, 225) puts it, "are distinctive because, much more than in any other sphere, non-functional causal relations govern social evolution."

It is a mistake to follow Alan Wilson's argument (1969, 229) that "we are more interested in the use of analogy by a geographical theory maker, and less with the philosophical arguments."

It is a mistake because analogy itself is a logical exercise and because coincidence does not mean the mere repetition of causality, which is, in any case, impossible.[15] As Ernst Mach put it (1906, 11), "there is always an arbitrary element in analogies, for they are concerned with the coincidences to which the attention is directed." The fragility of the method of analogy derives from the role it attributes to a priori and external factors. Reading the physical world onto the social world by analogy is therefore misguided.[16]

Often, the error is twofold. First, one of the principles of physical research is the search for ever greater totalities, on the basis of which the elements that make them up can be more effectively interpreted. Second, the idea that the physical sciences are *exact* disciplines is flawed. The representation that we make of the physical world changes in different periods in line with scientific progress. No truth about the physical world—let alone about the social world—is definitive.

Albert Einstein wrote that "the belief in an external world independent of the perceiving subject is the basis of all natural science" (1954, 266). We can extend Einstein's perspective to social sciences founded in objective reality. However, not all the claims of physics, not even of relativist physics, can be used in the same way to construct a theory or epistemology of the social sciences.

Its founders, full of zeal for giving geography a definitive scientific status, were mistaken when they believed the best route to achieve their aim was to construct the theory of a science of man on the basis of an analogy with the natural sciences. If it is absurd "to consider nature as itself a stranger to the spirit" (Husserl 1977, 8), it is also absurd to want "to erect the sciences of the spirit on the foundations of the sciences of nature, with the pretension of making them into exact sciences."

Possibilism vs. What?

Taken as a baseline of geography, the very dispute between determinists and possibilists has proved to be false.

The discussion of determinism was skewed from the beginning, starting from its own terminology. There was a confusion, whether deliberate or not, between the notion of *determinism* and what we might call *necessitism*. The first was used in place of the second, perhaps in order to discredit those who studied the development of history as the outcome of the complex, inexorable action of deep causes acting concretely and in concert at a given moment in time. In this famous dispute, the students of Vidal de La Blache have declared that they include human action as a factor, while the "determinists" (a name that the "possibilists" gave to Ratzel and his followers) assign an apparently self-evident causality to natural factors. This means overlooking the fact that not only do social determinations exist but that they affect man and nature equally. In any case, *determinations* are recognizable and measurable only a posteriori, and the idea of *necessitism* must be discarded.

In the preface of Lucien Febvre's book (1932, 11), the historian Henri Berr, when referring to what was then called "determinism," proposed *necessitism* instead. A determination, sociologically understood, must be clearly distinguished from a necessity. *Determinism is natural causality.* Among the causes in nature that *determine* phenomena, some are *contingent.* Among these contingent causes, some are geographical. The problem lies in knowing whether *geographical necessities* exist and whether natural phenomena can act as necessary causes on a "purely receptive" humanity.

Taken with its original meaning, the notion of determinism does not obliterate the idea of possibilism but reinforces it. When Vidal de La Blache wrote that "there are no necessities, but everywhere possibilities," he was stating an obvious truth. The reign of the possible is not the reign of the random but the conjunction of determinations in a given time and place. If we take the word as the possibilists meant it, we are not dealing with "fatalism" at all. Indeed, outside geography, before, during, and after this debate, the words "determination" and "determinism" have been used straightforwardly. This quarrel served only to slow down the evolution of geography and to mean that the notion of possibilism itself was never adequately developed.

2

Philosophical Inheritance

Eric Fischer (1969, 61) called Carl Ritter, Alexander von Humboldt, and Conrad Malte-Brun the "first modern geographers of stature." We might add Vidal de La Blache, Ratzel, and Jean Brunhes as "founders." They sought to identify laws and principles that would guide the nascent discipline of geography toward being a modern science.[1] Alongside the principle of the unity of the earth, Humboldt established the principle of general geography that Vidal took up. Ratzel laid out the principle of extension and Brunhes, of connection.[2]

It was, without doubt, progress. Though these ideas seem fuzzy today, their value as pioneering inspirations is clear.

Early nineteenth-century geographers were working before the emergence of the social sciences of Auguste Comte and Émile Durkheim. Their colleagues at the end of that century and at the beginning of the twentieth were influenced by historical events, but less so by progress in the social, natural, and exact sciences. Engaging in a terminological quarrel with the defenders of social morphology (proposed by Durkheim and his followers as key to a general science of societies) and ignoring the new concepts introduced by Einstein meant turning down the chance to progress more quickly toward a valid, total geographical theory. However, we can hardly criticize the founders of geography if even in our own time the lessons of Durkheim, Einstein, and others are not incorporated into geographical thought. An exaggerated obeisance to a closed circle of old ideas and threadbare refrains slows down our progress as a science.

We have to be careful when we refer to the philosophical inheritance of geography. What can any specialist, from any science, make of the ideas that are borrowed from philosophy? There are, without doubt, many paths to follow, of which one is the most extreme and another is the most curious. [On one hand,] one can naively apply random ideas without worrying about whether or not they fit. Karl Marx has been subject to this. He himself called his copyists "vulgar Marxists." Indeed, he went so far as to tell those who read his work and wanted to follow his method gently but firmly that they were not "marxists."

On the other hand, if we rigorously read philosophers appropriate to each particular field of scientific research and place them in light of present realities of the here and now, then we can establish general lessons. This is the right path, but it means that we must begin from the reality of real things, and not from concepts.

Others, who rightly or wrongly call themselves eclectic, take a little from here and a little from there, without the disciplining logic of the whole or concern for the compatibility of concepts. They make claims mechanically. Their sophistry and rhetoric are supported only by a formal logic, external to the reality in question.

It is not unfair to include many geographers in this last camp. Some change their philosophical position with the fashions. They are wind vanes who collect ideas as if they were choosing ties. Some are less naïve and develop their ideas not as pure science but to pursue external interests. Let us move on, however, to substantiate how such attitudes have led to confusion in geography and stopped it from holding an organized and enriching debate around a clearly expressed object. I will not preordain where this analysis will lead.

The Sources

To find the philosophical foundations of geographical science at the end of the nineteenth and the beginning of the twentieth centuries, we need to start with Descartes, Kant, Darwin, Comte and the positivists, but also Hegel and Marx.

Hegel influenced Ratzel and Ritter.[3] Marx influenced Ratzel, Vidal de La Blache, and Jean Brunhes. Yet, for a number of reasons, it was idealist and positivist legacies that in the end imposed themselves most powerfully on geography—that is to say, on official geography. Cartesianism, Comtism, and Kantianism were frequently added to the mix, along with the principles of Newton, Darwin, and Spencer.[4]

As Henri Poincaré (1905, 6) wrote, in respect of space and time: "it is not nature that imposes on us, but we who impose on nature."

A poorly constructed Darwinism pushed many geographers toward a determinism fed by a positivist ideal. Indeed, the fact that positivism even contaminated Marxism shows its importance in a crucial phase of the history of science. Jean Brunhes is an example of this marriage between Marxism and positivism, though Georgi Plekhanov has the seat of honor at that particular feast. An alliance of Marxism and positivism justified giving concepts from the natural sciences exaggerated significance. They influenced the human sciences by offering them the cherished laurel wreath of being *scientific*. Determinism drank from two springs: evolutionism and positivism.[5]

Though they did not say it explicitly, Vidal de La Blache and his school abandoned Darwinism and Spencerism. But it is difficult to lay bare their philosophical preferences and affiliations: they danced with Kant and embraced Marx but never abandoned Cartesian rationalism or the positivism of Comte and Poincaré.[6]

The influence of positivism reached Marxists themselves. Plekhanov is a good example of a determinist geographer: "in the final analysis, this structure (the structure of the collective) is determined by the properties of the geographical environment that furnishes men with lesser or greater scope to develop the productive forces" (Plekhanov, 1974, 711).

The positivists also draw on Newtonian philosophy. For Max Jammer (Plekhanov 1974, 98), "although Newton cannot ... be regarded as a positivist in the modern sense of the word, yet he drew a clear line of demarcation between science on the one hand and metaphysics on the other." Here, positivist thought is akin to Newton's ideas. It does not abolish metaphysics but separates metaphysical from physical investigation. In relation to space, however, Jammer argues (1969, 98) that Newton would have made an exception to his own rule. After all, it did not stop him from laying out a concept of absolute space. It is ironic to note that the philosopher and mathematician Poincaré (1914, 93) considered absolute space to be a "meaningless term." Newton himself fixed the idea of an absolute and unchanging space, of which relative space would be only a measure.

Up to a point, Kant confirmed Newton. In his own way, he revived Tomasso Campanella's idea of space as a receptacle. Newton and Kant are on the same side in the struggle between "possibilism" and "determinism," but they cannot help us make genuine advances in this debate. Regionalism is parallel to possibilism. It is determinism by another name.[7]

Little wonder, then, that the seemingly disparate notions of Kantianism and positivism are contemporaneous and that the Newtonian idea of absolute space resembles the Kantian idea of continental space.[8]

Each school connected to these philosophers defended some ideas, rejected others, and established its own path of action. Kantian propositions, whose influence Harvey (1969, 71) places a bit too late in the 1920s and 1930s, long helped European geographers combat the "determinist" current. Indeed, Harvey has clearly shown the relations between the regionalist idea and Kantian thought. In *Critique of Pure Reason* Kant proposes that space be understood as a condition of possibility for phenomena, and not as a determination resulting from them: "space is a necessary a priori representation that underlies all outer intuitions." Here we are much closer to the absolute space of Newton, space as "a receptacle." Space as "condition of possibility" ends up being the philosophical base as much for the possibilists as for the determinists. It is for precisely this reason that the possibilist school could never realize its aspirations.

But Newton in *Philosophiæ Naturalis Principia Mathematica* considered space to be *void* and empty, so Kantian space is also a "pure intuition" and not "a general concept of the relations between things." If we consider only *some* of their principles, we can take either Kant or Newton as a starting point and end up with the same results.

It is not surprising that the idea of time—the time of societies in movement—has been so absent from the minds of the founders of geographical science. Kant's space is tri-dimensional.[9] For Newton, time was a *continuum*, as absolute as space. The notion of time separated from space is responsible for the history–geography dualism that has elicited so many debates within and beyond advocates of interdisciplinarity. This binary still affects many geographers, ignoring how Leibniz's ideas of concrete time and space and their interrelations have been justified and renewed by Einstein.

Hegelianism and Marxism

The work of the founders of French geography is directly connected to Marx and the Marxists.

The notion of bilateral relations between city and region, appreciated by Vidal de La Blache, has dominated geographical work for a long time, principally though Raoul Blanchard and Georges Chabot and their pupils across the world in Canada, Latin America, Africa, and South and Southeast Asia. But this is not only a French phenomenon. Anglo-Saxon and German geography are full of similar approaches, even that of Walter Christaller, whom many consider a heterodox Marxist.[10]

We can speculate that interpretations of geographical evolution that attended the transition from the Middle Ages to the capitalist era were more influential than they should have been. But times changed. Writing in 1857 in the *German Ideology,* Marx referred precisely to that period of transition: "the city, with the territory that surrounds it, forms the economic totality." This concept, though useful to describe conditions of spatial organization at the end of feudalism and the beginning of capitalism, is less applicable to other moments. By following Marx too closely, the founders of geographical science made an interpretative error. It persists today. A specialist in regional studies recently wrote that "the growth of a central place must be sustained by its region." For this noted specialist (Richardson 1969, 106), "the most striking contrast emerges from the fact that if the growth of a central place is underpinned by its complementary region, the growth of its region of influence is maintained, in the first place, by the pole."[11] As I have already sought to demonstrate in my books of 1971 and 1975, this framing defies existing reality. What, today, is the "complementary region"? How do we define the "zone of influence"? Marx's temporally specific idea about a particular period is also resuscitated—though perhaps not intentionally—by Brian Berry and John Friedmann.

Plekhanov (1967, 40) first called attention to the similarity between Ratzel's language and Marxists'. For instance, Ratzel wrote in *Volkerkunde* (1887, I. Band, s56, English version trans A. J. Butler 1896, 26): "the great problem is not to make the purchase of food easy, but the fact that certain inclinations, habits and, ultimately, necessities, are imposed on men." In the same work (I. Auflage, s.17) he continued:

The sum of the acquirements of civilization in every stage and in every race is composed of material and intellectual possessions. . . . They are not acquired with like means or with equal ease, nor simultaneously. The material lies at the base of the intellectual. Intellectual creations come as the luxury after bodily needs are satisfied. Every question, therefore, as to the origin of civilization resolves itself into the question: what favours the development of its material foundations?

This is very similar to Marx's conceptualization of superstructure and the connection between material elements and the facts of production. For Plekhanov (1974, 77), Ratzel's is "a clear historical materialism, even if it does not have the same qualities as the materialism of Marx and Engels."

What about Jean Brunhes? Upon reading his work, it is astonishing to note the similarities of so many of his formulations with Marxist ideas. The surprise is lessened only by the fact, already noted, that in their enthusiasm to characterize geography as a science, some of its founders were attracted to positivism, in which they found both inspiration and reassurance. It is perfectly possible that Marxist positivism influenced Jean Brunhes. He was fourteen when Marx died. His book on human geography has the subtitle "A Positive Classification."

One of Brunhes's concerns was precisely to make a positive classification of geographical "facts" or elements. He put them into three large categories: productive, unproductive, and destructive. The productive elements of the occupation of the soil include domestification, agriculture, and pastoralism. The unproductive elements include houses and settlements and means of transport and communication. The destructive elements are mineral exploitation and the killing of plants and animals. Houses and roads, wrote Brunhes (1956, 28), "are interconnected, and linked to the inhabited earth; they represent the two human facts that can legitimately be called, in a positive sense, and without giving the word 'unproductive' a pejorative connotation, 'the unproductive use of the soil.'"

I am conscious of not stretching the point about analogies, but in this key passage I hear an echo from elsewhere. In *The German Ideology* (1947, 69), Marx wrote that "in the development of productive forces there comes a stage when productive forces and means of intercourse are brought into being which, under the existing relationships, only cause mischief and are no longer productive but destructive forces." In the *Introduction of 1857,* Marx added: "a railway on which no one travels, which is therefore not used up, not consumed, is only a railway potentially, not in reality." To return to Brunhes, this is an example of a geographical element that is *unproductive,* a road not dynamically connected to houses.

Edward Ullman (1950, 31) puts the problem in similar terms: "this does not mean that transport develops automatically. We are dealing with a passive force, a necessary but not a sufficient condition, however profound its effects on spatial organization."[12] Ullman is an American geographer who is able to carve out new historical trajectories and whose thought owes much to the European founders. However, this does not automatically connect Ullman to Marxism. Otherwise, we would have to put Brian Berry in the same company when he is speaking in his fluid way of "systems of systems (etc) of cities." Here the word "system," even though followed by the preposition "of," could be reproduced as many times as you like. Is this not the same idea, to some extent, that can be found (with the help of some physical concepts) in chapter 13 of the first volume of *Capital*? Still better, Berry (1964a, 3) speaks of a geography whose integrative concepts and process concern the global ecosystem of which man is a dominant part. This approximates Marx when he writes that nature and man form a unity, such that man is a part of the nature that he himself modifies. However, this did not stop Berry from formulating a notion of systems of systems (of systems . . .) of cities, without taking into account the notion of the totality. As in so many other formulations of this type, the global ecosystem appears in the description but, as if by magic, disappears in the interpretation. This is not an ecosystem with a universal foundation but a return to a stiffer regionalism.

Let us return to the founders, though, and leave the living in peace.

In an article by Vidal de La Blache (1899, 106), we find the following ideas: "A people, however primitive, leaves its mark upon the objects they make, whose substance and forms are taken from nature. These objects belong to that particular people." Here is a Marxist theory of the unitary relations between man and nature in the writing of a non-Marxist geographer. The nature of which Vidal de La Blache speaks is humanized nature, and the substance taken from it to make objects is human labor.

However, in putting forward a dualist and reductionist notion of regional geography, Vidal de La Blache did not follow this reasoning. He, like other geographers of his generation, sought to define the very particular relationships that connect man and the space that surrounds him. For example, in the idea of *genres de vie* the personality of man ends up defined by a regional personality. It is as if deep down he was aligned with Marx but ended up distorting reality. Writing at the high point of industrialization and imperialism, he ignored the reality of the economic and social division of labor. By then it should no longer have been feasible for such methodological mystification to reestablish a long-surpassed account. At that time, indeed, nothing that took place in the regions of France, or of any other European country, did so without a direct or indirect relationship with national and global economic events.

From Descartes to Total Eclecticism

The enduring influence of Descartes on geography, as on other scientific domains, is evident in the developed world. The search for rational knowledge, resulting from a sui generis dialectic that distinguishes pairs of categories capable of constructing an indissoluble but not contradictory unity leads, in geographical questions, to the justification of a distinction—even a disjunction—between a general geography and a regional geography as the inverse of each other. They end up opposing each other.

Regional geography, defined zealously as a search for the "concrete," rests on a notion of abstract, nonrelational space. General geography, constructed on the basis of principles, does not bother itself with the historicization of concepts. It is condemned to become a theoretical endeavor without an epistemological undertaking; a useless attempt with no consequences.

Albert Demangeon is a good example of the incapacity to philosophically align general geography and regional geography. Initially, he showed complete loyalty to the idea of the totality, the *unity of the earth*, as his concerns with international economy demonstrated (for example in two classic articles in the *Annales de Geographie* in 1929). When, however, he formally outlined his method in the introduction to his posthumous *Treatise on Human Geography*, while he alludes to "general facts" (1947, 25–34), he remains loyal to the principles of classical regional geography. For him, regional geography constitutes "one of the key supports of general geography. . . . It is wise to begin from the particular, the local and the regional, and observe what the region contains in its contours, its plants, its inhabitants, and also to explain how this dynamic thing results from the union between a fragment of the earth and a group of human beings."

This long, elegant sentence makes no reference to the conditions of international economy that Demangeon studied, nor to laws whose scale exceeds the dimensions of a place and that underpin the relations between a fraction of humanity and a fragment of nature.

This approach can be explained by the philosophical eclecticism that has directed geography from its infancy as a science. It has paralyzed the development of the discipline and annulled the serious and well-intentioned efforts of its founders and so many of their followers. The philosophy of geography, whatever direction it may take, cannot continue to be stitched together like a patchwork quilt.

3

Postwar Renovation

"A New Geography"

Geography is not immune to the huge transformations in all scientific domains after the Second World War.[1] In the human sciences, this has been a story of revolution, not evolution. Three essential reasons underpin it: first, the foundations of scientific work have significantly advanced; second, the needs of its users have changed; and, finally, the object of scientific activity has changed.[2]

Researchers have new methods, tools, and technologies for approaching reality at their fingertips. At least superficially, this has given research the means to define realities with greater precision and to put forward new laws for debate.

This has led scientific activity, including geography, in new directions. It is impossible not to be struck by the diversity and novelty of recent geographical work from all over the world. The breadth of approaches must certainly seem unfamiliar to readers used to the specialized journals of before 1950.

We often hear talk, therefore, of a "New Geography" that characterizes itself as different, in opposition and even in contradiction to a "traditional" geography. The choice of terms is deliberate. The protagonists of this new approach have sought to distance themselves from a geography that they see themselves as having not only superseded but as having been a "non-geography."[3]

The new approaches have had differing appeals in different countries. In some places, people have remained attached to old or new principles, methods, and modes of working and have sought to perfect them.

Bolstered by new methods, others have aligned themselves with the "new paradigms." They have often ended up using a lexicon distinct from that of traditional geography, deploying accessible language to present facts and ideas elegantly.

"Traditional" geography was forged under the influence of "national schools." However, from the 1950s and 1960s on, we find a methodological school that aims to overcome local exceptionalism and that establishes organizations and publications of its own. This school disseminates its ideas through conferences, workshops, and academic exchanges that transcend national boundaries. The center of this tendency is the Anglo-Saxon world, but its access to powerful means of communication make its reach international. Geography has thus reproduced the modalities of economics and politics, which have globalized to a previously inconceivable scale.

Crucially, and polemically, the expression "New Geography" asserts that the new is also the unique. Depending on the particular political conditions of each country, this terminology and its connotations have provoked reactions from indifference to perplexity to an antagonism toward extremists of both poles. Some affirm the need for the new tendency (and the new denomination), and others deny it absolutely. Between the extremes we can find a number of intermediate positions.[4]

Geography is facing a battle like that out of which it emerged as a science at the end of the nineteenth century. On one side is a cold and pragmatic quantitative trend and on the other a more speculative, social disposition. The concerns of writers such as Maurice Le Lannou have found an appreciative echo both inside and outside France. The enormous systematizing endeavor of Maximilien Sorre has been equally influential. Sorre, though, has not cultivated a set of followers, so his work has not become a "school," nor has it been disseminated as widely as it might otherwise have been. Pierre George, meanwhile—like Carl Sauer and Richard Hartshorne in the United States—has found his fruitful work multiplied and diversified by his students. A neo-Marxist tendency also emerged over the end of the 1940s and the beginning of the 1950s. What this tendency represented, and the difficulties that it raised, was the subject of my 1975 article in the North American journal *Antipode*.[5] Indeed the older tendencies (such as the regionalist disposition in France) were still powerful and, faced with a New Geography, seemed to accrue a vigor that coming under attack often lends to ideas. Indeed, the most criticized tendencies won an unfortunate victory that prevented more lucid approaches from reaching their ultimate conclusions. All ended up trapped by an ecological narrowness (or a narrow ecology), working with a truncated totality that valued the "not real."

This is the common ground of all the old tendencies, their successors, and the new, so-called revolutionary fields. Thinking itself to be challenging the New Geography, traditional geography ended up helping it, nipping in the bud any chance of a renovation from within. So did those who merely surrendered to whatever trend was in the ascendancy.

The debate is far from over. Isolated voices questioned the destiny of geography before the end of the 1960s, when disenchantment with the quantitative revolution began to be felt.

This is a serious question. It relates to the present and future of the discipline. We should, first, affirm the existence of the so-called New Geography, because it is not completely dead. We should, next, understand what it consists of, its goals, its optics, and its methods. We should understand what its object and objectives are before coming to terms with its fundamental weaknesses.

The so-called New Geography took shape above all through quantification. But it has also made use of models, systems theory (including ecosystems), the thesis of the diffusion of innovation, ideas about perception and behavior, and various approaches to empirical and ideological evaluation. In the coming chapters we will lay out a succinct outline of its key aspects before appraising their "thingism" and ideologism.

4

Quantitative Geography

In 1963, Ian Burton argued that the quantitative revolution had made our discipline into a respectable science.

The search for a mathematical language in geography emerged from geography's quest, in different guises at different moments, to be scientific. Mathematical methods are considered the most precise, the most generalizable, and the most predictive, thanks to the combination of systems analysis, modeling, and statistics.[1] The quantitative approach responded to a desire for rigor, in which cause and effect are identified through linear, predictive, and historical models. For quantitative geographers, multifactorial analysis will once and for all resolve the complexities raised by the breadth of variables at stake. It will overcome the previously insuperable fragility of interdisciplinary work.

The possibility of separating variables is at the base of quantitative work. Explanation is strengthened as this approach can not only understand differentiation but also account for it. It can build descriptive and prospective models. This capacity for prediction is not intuitive or emotional but systematic.

Looking to the exact sciences for how to smoothly apply quantitative methods reflects a constant desire for measurement. Systems analysis and the construction of models are applied to the study of space precisely in order to understand and define multiple variables. Indeed, they are the cause and effect of the famous "quantitative revolution." Diffusionist theory, as Hägerstrand himself (1967a) argued, also relied on quantification.[2]

Quantification in Geography

For Chisholm (1975, 26), the roots of quantification in geography are not in modern statistics but in cartography. He was dealing with a different kind of quantification, in relation to a specific kind of geography, "used to secure a more exact description not, as in the modern period of quantification, oriented towards explanation within a probabilistic framework."

The case for the quantitative method has been outlined very unevenly. The English geographer Alan Wilson was, one hopes, being ironic when he wrote (1969, 230) that "the theoretical geographer"—by which he means the quantitative geographer—"does not need to be an *original* mathematician or statistician." Michael Hurst writes that "in fact it is now comparatively easy in geography to describe fairly complex patterns in mathematic terms without any understanding of the basic processes involved, as in simulating the diffusion of an innovation through space, without understanding why some people accept innovations and others do not."

Linearity, Colinearity, Etc.

The search for causality and linearity is a preoccupation for those who use quantitative methods in geography. Right at the beginning of his article on migration patterns, Barry Riddell notes that regression models "have been among the most commonly employed tools in the search for understanding complex, multidimensional spatial processes." For him "the assumptions . . . of linearity, normality and multi-collinearity of the model are basic to the estimation of the parameters" (1970, 403). But the fundamental hypothesis is flawed. A multidimensional process cannot be contained in a linear model because we are dealing not with relations of cause and effect but with a network of causality at different levels: what might better be called a "context." Not doing so means working with "independent" variables, as Riddell himself did in the African example that serves as the basis of his thesis. In his study of Sierra Leone he begins from a priori hypotheses instead of from reality itself. His point of arrival is, predictably, an exercise in abstract empiricism, whose value for concrete knowledge of concrete reality is meager.

Douglas Amedeo and R. D. Golledge suggest "the possibility of non-linear relations" through examples that include interactions described as "*exponential*" relations:

> suppose we have two sets of numbers. In each of these sets there are correspondences in that, associated with each number in the first set, there is a corresponding number in the second set. The relationship that specifies the nature of the correspondence between the exact numbers in each set is called the *functional relationship*. Returning again to our two number sets, we refer to the first set as the *domain* of the function; the second set therefore would constitute the *range* of the function. (1975, 82)

This alignment of correspondence, far from overcoming linearity, only multiplies it.

The same thing is put differently by Sylvie Rimbert when discussing methods of analyzing multiple variables in geography:

> it has been argued that geography was a science of the relations between multiple variables observed in the landscape. These relations can be specified through various inductive statistical methods that associate the *variables initially two by two,* and later in larger numbers.
>
> The connection that can pertain between the series of values of a variable and the series of another variable is expressed by a certain degree of correlation that is generally calculated in two ways: *the Spearman coefficient* (1905) for pairs of variables or *the Pearson coefficient* for measurable pairs of variables.[3] These two coefficients have values between +1 and −1. Once calculated, the correlating coefficients for a great number of pairs of variables can be classified in a framework called a *correlation matrix* that, in some cases, can become a *matrix of data for factorial analysis.* This last operation consists of substituting one framework of coefficients for another, simpler one, in which only a limited number of *independent factors* explain the existing connections between diverse variables. (1972, 103)

In his commentary on Tinbergen's method, Keynes asks:

> Am I right in thinking that the method of multiple correlation analysis essentially depends on the economist having furnished, not merely a list of

the significant causes, which is correct so far as it goes, but a *complete* list? (1939, 560)

Herein lies the difficulty (and the weakness) of factorial analysis, in spite of the enthusiasm that the method has long elicited. Harvey (1969, 343) provides lists of geographers interested in this subject and works that deploy it. Michael McNulty (1969, 343) reminds us that among the first studies that used it were some that took English cities as their object, with the "objective ... to assemble and collate material, pointing out the similarities and contrasts, and secondly to classify towns on the basis of their social, economic and demographic characteristics."[4]

This widely followed process was later abandoned. Brian Berry disavowed it in his chapter in *Directions in Geography,* edited by Richard Chorley.

Measuring to Reflect, or Reflecting to Measure?

Even before the current spirited debate, Sorre (1952, vol. 2) wrote, deploying a Spinozan phrase, that geography was "a meditation on life, not on death": "death was inflicted by superficiality, by merely formal description, by statistics set out merely for the pleasure of manipulating numbers, and by modes of classification which seek only to lock reality up."

Criticizing the use of quantification in biology, Henri Bergson, cited in D'Arcy Thompson (1917, 721), argues that "calculation touches, at most, certain phenomena of organic destruction. Organic creation, on the contrary, the evolutionary phenomena which properly constitute life, we cannot in any way subject to a mathematical treatment."

Alfred North Whitehead (1948, 127) condemns the "new modes of error" that mathematics can lead to, principally because it introduces "the doctrine of form, stripped of life and movement." Hence the affirmation of E. J. Bitsakis (1974, 31), according to which mathematics is "an abstract and mediated reflection of the real."

Among economists, the use of quantitative methods has been frequently contested. The Mexican Alonso Aguilar critiqued them in his book *Economía política y lucha social,* while Aníbal Pinto and Osvaldo Sunkel (1966, 83) have argued that "the use of mathematic methods is not the only route towards scientific rigor."[5] Peter Bauer and Basil Yamey (1957, 14) are less peremptory. For them, "the difficulty or impossibility of submitting certain economic phenomena to quantitative treatment . . . does not mean that the phenomena cannot be analyzed" and "quantifying something does not necessarily show us its most important aspects."

The abuse of statistics has also been criticized. Armand Cuvillier (1953, 165), a sociologist, reminds us that "neither an accumulation of brute data, nor a simple register of particular facts constitutes a science."

A futurologist, Andrew Shonfield (1969, 26), suggests that "straightforward statistical measurements acquire significance only when the speculative social imagination is applied to them."

In general terms, quantification is critiqued by philosophers. I have already cited Whitehead, but there are many others. As Gaston Bachelard (1947, 213) wrote: "It is necessary to reflect in order to measure, and not measure in order to reflect."

The Problems of the Quantitative Approach

Burton (1963, 151–62) puts the adversaries of quantitative geography into five groups. The first are geographers who rejected the "quantitative revolution" from the outset, thinking it would lead geography down the wrong path. The second group includes geographers who thought maps were sufficient to express the interactions that characterize the organization of space. A third group argued that "statistical techniques are suitable for some kinds of geography, but not all geography" (155). Another set of objections is more measured: quantitative techniques are desirable, but the number of errors in their application militates against them. The last group prefers to bring to bear more personal criticisms: for them quantification would be a good thing, but quantitative geographers are themselves no good.

But there are more serious critiques to be made of quantitative geography.[6]

Paradigm or Method?

Is quantitative geography a paradigm or a method? "Theoretical geography," or "geographical theory," calls itself a new paradigm of "locational" study. Emerging in part from planning, it enthusiastically uses new theoretical approaches such as systems analysis and modeling and makes projections and predictions.

Quantitative geography, however, would constitute a methodology, or the beginning of a paradigm, only if it were underpinned by theoretical approaches. Is there a necessary association between a paradigm and a method? We must approach this question from two angles. First, could the concern with quantification have come before quantitative geography itself? Yes. Geographers have always sought to back up their affirmations with both statistics and research. H. C. Brookfield argues (1964, 300) that "much of the excellent work now emerging from the application of mathematics to the analysis of distributions is merely a refinement and sophistication of geographical description." The novelty is the use of modern mathematics not only to analyze data but to collect it and use it as a means of expressing results.

The second question is whether the new paradigm could be achieved through quantitative geography. The quantitative—or merely statistical—contribution is relatively useless, even toxic, without a systematic understanding of the tools it relies on.

But the fact that both paradigm and method emerged at the same time, and in parallel, can lead directly to improving the methods we work with, while at the same time improving our ideas and theory. For geographical progress we must avoid this. Edward L. Ullman (1973, 272) flagged this problem clearly when he argued that it was a mistake to make the quantitative method synonymous with spatial analysis. "Quantitative methods," he wrote, "can be applied to most approaches to geography, and by themselves do not constitute a geography. One might argue that they are a desirable, but not a sufficient condition."

The obsession with quantification and measurement encouraged geographers such as Timms (1965, 239) to affirm that without it and its precise, objective explanations, comparison and abstraction become impossible. As Philip Stone et al. (1966) explains, validity in content analysis is the "degree to which a measuring instrument is capable of achieving the aims for which it was constructed" (see Moodie 1971, 148). We take our point of departure from the tool of measurement, not from that which is to be measured. The privilege given to methods and techniques is one of the most serious weaknesses of so-called theoretical geography.[7] It is not difficult to fall into the tendency that Norton Ginsburg (1973, 2) critiques when he writes that "theoretical inquiry becomes concerned primarily with those subjects that are more or less readily known, to which the application of available technique is most convenient."[8] The great mistake of so-called quantitative geography was to treat a method—indeed, a questionable method—as a theory.

Using the expression "quantitative geography" to describe a new geography introduces doubt and confusion. The terms "mathematical" or "quantitative" geography can be applied to a number of geographical paradigms, new and old, including some that are obsolete. Quantification is just an instrument or, at most, the principle instrument. It would be better to call attention to theoretical or conceptual aspects, that is, to paradigms themselves. The construction of theory remains fundamental.

There is no real opposition between the quantitative and the qualitative. Some wish to argue otherwise, but it is clear. Everything presented as quantitative is the numerical transcription of a fact or a prediction based on a sequence. The struggle to separate particular variables for a given issue relates more to questions of theory than to questions of calculation.

This is when the problem becomes most acute. Greater or lesser ability to distinguish between variables determines the success of quantitative analysis. This takes us to a much more general question. Analysis of geographical realities is invalid without a theoretical armory capable of recognizing the intrinsic value of each variable.

To distinguish between significant variables, we need to define them clearly. This definition, this choice, is undertaken within the framework of a value judgment, a theoretical position, and in terms of concrete reality in movement. The qualitative, therefore, takes precedence. Once a choice is made, we can move to the second stage of accounting for phenomena. This is crucial if we are to present rigorous results or refine theories.

To work in the other direction is to suppress the effort to consider different explanations and so effectively to rule them out. This would repeat the errors of the past. Berry and Pred (1965) recognize the necessity of concepts when dealing with the use of quantitative methods. But it is one thing to start from concepts based on concrete reality and another thing altogether to apply a hackneyed and ideological epistemology in which parameters identify their legitimacy in relation to other parameters and not in relation to the objective appearance and conjuncture of things and events. As Broek (1967, 50, 105) points out, the quantitative approach leads us to construct abstract models.[9] In a similar vein, Hurst (1973, 46) notes that the majority of what we experience in the landscape cannot be subject to quantitative analysis.

The Greatest Sin

The greatest sin, however, of so-called quantitative geography is its total denial of the existence of time and its essential qualities. The current application of mathematics to geography allows us to work with successive stages of spatial evolution but is incapable of saying anything about what happens between one stage and another. Therefore, we witness the reproduction of successive stages but never their succession. We are working with *results*, but *processes* are omitted. Such results lead not to interpretation but only to mystification.

Can we know a thing without knowing its genesis? The space that mathematical geography reproduces is not the space of societies in movement but a snapshot of some of its moments. Photographs can only describe, and description should never be confused with explanation. Only *explanation* is scientific.

5

Models and Systems
The Ecosystems

Analyzing Systems

Systems analysis has contributed significantly to the progress of the exact sciences and has been used in the human sciences for at least twenty years.[1] Geography is one of the last to deploy this method.[2]

Space, the essential object of geographical studies, when considered as a system, independent of its dimensions, could be susceptible to systems analysis. This would construe a hierarchy between different spaces and systems and contribute to explaining localizations and polarizations.

For Chisholm (1967), geographers have already studied space in terms of systems, although they used different terms. He mentions, for instance, cycles of erosion and functional regions. Another geographer, Rodman (1973, 100–105), shows how this approach was already familiar in the Soviet Union in spite of the expression "territorial system" being recent.[3]

Cities and urban networks have also been considered in terms of systems. For Berry (1964b, 148) "urban theory therefore may be viewed as one aspect of general theory of systems." For Richard Meier, similarly, "the city is a complex living system. Its anatomy and composition can be studied and analyzed like any other living system" (1962, 1).

In his classic article "Cities as Systems within Systems of Cities," Berry wrote: "the previous findings point in one direction: that cities and sets of cities are *systems* susceptible of the same kinds of analysis as other systems, and characterised by the same generalizations, constructs, and models" (1964b, 158).

In this respect Harvey (1969, 453) observes, ironically:

42

the notion of systems embedded within systems within systems *ad infinitum* is an attractive one. It poses no mathematical difficulty for we can simply group elements into a hierarchy of "classes" with each higher-order class forming an element in a higher-order system.

In Fred Luckermann's opinion, "the geographer can conceive of points on earth as parts of a system, and one relates to another, according to different levels of interaction (in Abler, Adams, and Gould 1971, 54).[4] Still, in the analysis of systems, the geographical fact within the definition of the "element," used elsewhere by Harvey (1969, 452), "from the mathematical point of view an element is a primitive term that has no definition." Thus, says Harvey, "mathematical systems analysis can thus proceed without further consideration of the nature of elements." This leads directly to a tautology: "but the use of mathematical theory of systems to tackle substantive problems depends entirely on our ability to conceptualize phenomena in such a way that we may treat them as elements in a mathematical system" (Harvey 1969, 452).

It is a dead end.

A system is defined by a center, an edge, and the energy that connects them. The original characteristics made in the center project themselves to the edge, which is modified by them.

It is only on this basis that we can systematically understand the articulations of space and begin to distinguish their nature. This enables us to designate each piece of earth in a precise and particular way. Each spatial system and its corresponding localizations appear, therefore, as the outcome of a play of relations. Our analysis will be more rigorous if we avoid setting up confrontations between simple variables as precursors to causal analyses. Doing so artificially isolates certain variables and limits understanding of the totality of interactions.

Because the spatial system is always the consequence of a projection of various historical systems, one system is always substituted for another. As space contains the characteristics of variables of different ages, such a focus should enable a more cautious and systematic interpretation of what survives and how things are articulated.

The problems of the relations between what is current and what is past would be much easier to resolve if they were studied beyond the limited framework of the particular histories of each variable. In effect, the evolution of space is not the result of the sum of the histories of each given thing but the result of a succession of systems.

If we start from this perspective, the problem of the scale of study accrues a new dimension. If, for the purposes of analysis, it is possible to delimit a certain part of space, it does not follow that analysis is then circumscribed to this geographical scale. On the contrary, the scale of study surpasses this "natural" scale whenever variables are defined in relation to larger systems.

The Ecosystems

Among the new tendencies is that of frequently addressing space in terms of ecosystems.[5]

We might imagine this, at first glance, as a return to an older orientation of human ecology in the United States. We can associate human ecology with the European school of regional geography.

The two approaches are very similar. Regional geography studies spatial differentiation mediated by the relations between nature and human societies. Human ecology is concerned with the forms of human adaptation to different environments and the material consequences that arise from them.

The idea of the ecosystem revives scientific propositions, but human ecology's underlying methodology is different—it goes beyond the study of natural facts. Certainly, the notion of ecosystem applied to the explanation of space emerges partly from progress in the disciplines of natural ecology. However, while there is a methodological affiliation, the content is broader.

The notion of ecosystem should enable historical and natural subsystems to be incorporated into spatial analysis: first, in terms of how human societies use natural conditions in different historical periods, and second, in terms of how nature itself is transformed by men. As history goes on, successive human groups form relations with a natural framework that is already modified by men.[6]

If space cannot be defined by the bilateral relations between man and nature, neither is it the exclusive result of economic changes, as if the surface of the earth was a smooth field of action and not a rough and contoured terrain. To its credit, human ecology has overcome objections against economistic geographical approaches based only on the principle of location.

The great difficulty of a regional ecology approach is the impossibility of delimiting the totality of economic, social, and political phenomena to a particular area, when their scale of action surpasses their place of manifestation. Whenever there is a disjunction between these two givens, regional geography risks becoming a thin description, merely a study of aspects.

Systems and Quantification

Reino Ajo (1962) shows the connections between systems analysis and mathematical models. In his article "An Approach to Demographic Systems Analysis," he writes that it is only through a mathematical understanding of the equations that govern a system that we can understand the specificities of its functioning.

According to V. Y. Vagagini and G. Dematteis (1976, 126), one of the great weaknesses of systems analysis is that, in the service of the "quantitative analytical method," it cannot consider and analyze the historical relations of form—what they call the "territorial structure"—alongside processes.

To do so requires a method that can account for the qualities of variables themselves and their propensity to combine in specific ways under specific temporal and spatial conditions.

Though not mutually exclusive, it is in this sense that systems analysis, modeling, and quantitative approaches are somewhat anemic.

The use of mathematical models in association with systems analysis has led to some key observations. One of these comes from Gunnar Olsson (1967), for whom "similar mathematical formulas can be applied, unproblematically, to completely different phenomena."

In 1974 I wrote:

> To consider space as a system, a generally accepted approach, is not sufficient. We must know how to define a system. If we content ourselves with the classical definition in which the system is a set of elements, the relations between these elements, and their respective attributes (Hall and Fagen 1956,

18), we will struggle to achieve an operational definition of space. In fact, as Maurice Godelier puts it (1972, 258), "a system is a group of structures interconnected by certain rules, it is the structures that are defined by a group of objects, interconnected by certain rules." (Santos 1974b, 273)

The use of "system" as a synonym for "totality," implicit in Montesquieu's work, is explicit elsewhere.[7] For Marx, the definition of system is not so far from that of structure and totality. According to Maurice Godelier's (1966a, 829) explanation, "a system ... is a determined combination of specific modes of production, circulation, distribution and consumption of material goods." The social totality is defined based on a system—the economic system—and a structure: the economic structure.

Marta Harnecker (1973, 84) makes a distinction between these two concepts when she writes that "the economic structure [is] the set of relations of production" and "the economic system is the whole economic process: production, distribution, circulation, consumption."

Indeed, the economic structure is the combination of the mode of production and the superstructure. According to J. L. Ceceña (1970, 168) "the mode of production is, for its part, the unity of productive forces and the relations of production, the set of which we can call the economic base or the infrastructure." As Chisholm argues (1966, 221), all parts of the economic system are interdependent and, consequently, there will be ramifications if any point within the system changes.

When K. Boulding (1966, 108) argues that, of all disciplines, geography is "the one that has caught the vision of the study of the earth as a total system," we should take this less as a compliment than as a statement of intention. In fact, the understanding of space as global space is insufficient if we do not think of society in similarly holistic terms. It is possible to think of space as a system and yet take into account only the relations between spatial objects and not social relations.

As it is interested in how sections of reality interact, systems analysis contributes to a holistic knowledge of reality. But it is fatal to consider parts in relation only to one another, as if their movement does not effect the totality. To put this in spatial terms, it is as if the relations between New England and Texas had no repercussions for the United States as a whole and were not conditioned by the nation as a whole. The fault lies in reorganizing the reality being analyzed, instead of reproducing it. Of course, there is a lot to be said about the so-called relations between parts of space. But real knowledge of space is given not by relations but by processes. Systems analysis neglects this because it was created for, and applied to, mathematical models. These mathematical models, when they refer to space, have a fundamental weakness: they cannot interpret time and the movement of time. Because when we talk about process, we are talking about time.

Models in Geography

The difference between a system and a model is much more than terminological.[8] In each, the model is defined in two ways. On the one hand, it can define a set of local systems in a given historical moment acting in different sites within the same space.[9] On the other hand, models can simulate the evolution across time of local systems, each understood as the result of another *local* system. The first would be a descriptive model, the second an evolutionary model. Predictive models incorporate both evolutionary and descriptive qualities in order to understand vertical and horizontal dynamics, the totality of mechanisms and trends without which models are neither possible nor predictive.

Models are not necessarily interpretative; they can be purely descriptive. This does not mean they do not need to be inscribed in a theoretical framework. Getting valid results from any investigation depends on this.

Interpretations diverge when we discuss the influence of cities on a region and affirm that in underdeveloped countries the great urban agglomerations "absorb" their immediate surrounding space. Some see the city as the cause of this absorption, others as a negative disequilibrium, yet others see a trace of unity only in the city, such that another external pole, at a greater scale and endowed with a directional force, imposes itself as much on the city as on its region.

The same happens in terms of the spatial repercussions of growing populations in particular countries; what used to be known as "demographic pressure." Either we can conceive of demographic pressure as simply the direct consequence of demographic growth, or we can take into account the distribution of wealth in global society: the income of one part of the population—the more numerous part—is not sufficient to supply its needs.

We can put the question of *favelas* alongside demography. The existence of *favelas* in cities in underdeveloped countries is often seen as the result, on the one hand, of demographic expansion and, on the other, of the lack of dynamism of cities that are incapable of providing the requisite number of jobs. But in order to interpret the phenomenon of the *favelas* we should begin from a different optic and focus on the irresistible attraction to the masses of new types of consumption in cities. In fact, under contemporary conditions of public health, new products bought with cash or with easily available credit offer people elements of comfort and prestige. They are considered indispensable and are often chosen even above the search for a decent dwelling.

Underlying so many urban phenomena is the problem of employment. It is at once a problem of urban morphology and of urban economics. For many, this problem of employment originates in the imbalance between the number of posts offered and the uncontrollable mass of applicants, including migrants. However, we can interpret it differently, as the adaptation of the urban economy to the imperatives of imported technology when the state itself does not have the means to underwrite, either for the city or the country, a political economy that would drive the creation of a greater number of permanent jobs.

It is starting from here that an interpretation of concrete reality emerges. The importance of a theoretical framework is obvious. The methods that organize reality into particular schema are themselves merely secondary instruments.

The Construction and Success of Models

There are two approaches to constructing models.[10]

The first is to make the model ever more complicated. This requires using a vast number of variables to take regional and local specificities into account. But the elaboration of the model and the multiplication of its terms can lead to it becoming complex and unmanageable. It risks becoming an "anti-model."[11]

The second is to base local or regional models on simple general models, to which local and regional parameters and variables are added.

The results differ depending on which basic approach we use. In the first case, improving the model implies an inductive approach. We can enrich a model by improving its rationale. As it is the method—the instrument—that determines the approach, we can end up more worried about external facts than about the reality being analyzed.

In the second case, improving the general model implies a deductive approach. It is from inside reality that the general model is enriched or contradicted. But if a general evolutionary model is applied to contemporary theoretical cases or to contemporary descriptive models, the particularities of each country have to be considered. This suggests accounting for distinct histories and adapting the periodization or chronological subsystems of general models. Local data of all kinds—natural, cultural, economic, political, and so on—would need to be introduced to make external inputs autonomous. To analyze the interior of a space in the contemporary moment involves studying its diverse subsystems, local forms, export practices, governmental conditions, and more.

In fact, each of these subsystems behaves differently, with different degrees of interdependence, impacts in space, relations with employment levels, and so on.

If we approach the functioning of the contemporary situation, the intermediation of evolving variables, and the successive spatial forms of each subsystem, then these two lines of enquiry are worthwhile because the analytical method enables us to reconstruct the totality.

The greatest mistakes in the use of modeling in geography are in the mechanical practices—aggravated by the use and abuse of quantitative geography—by which concepts are transformed into metaphysical categories and history is frozen into fixed schemas. A model is a representation of reality whose use is justified only if it helps us understand reality, that is, if it serves as a hypothesis for further verifiable work. Just as empirical facts become theory through the intermediation of historical concepts and categories, we can get from theory to the empirical thing through models. We can put theory to the test in this way—with or without the intent of reformulating it—because reality is mutable. The model is thus employed at the same level as the concept in a constant process of to-ing and fro-ing from raw facts, to theory, and back again to the empirical thing.

This movement means that facts can be better understood by the use of theory and that theory can be improved by the proof of facts.

Both concept and model must, therefore, be permanently reinterpreted and remade. This can be achieved only if we recognize that theory, like reality, is in a permanent process of evolution.

If we forget all this and apply a frozen model to explain a reality in motion, then we are engaging in a straightforward methodological violence that leads not to scientific authenticity but to error.

6

The Geography of Perception and Behavior

The geography of perception and behavior is a new trend within the discipline. It owes a lot to psychology and social psychology.

It is based on the idea that each individual has a specific way of apprehending and evaluating space. Geographers like Raymond Ledrut (1973) have sought to define a type of social space for each individual, in the city and beyond. This social space can be defined by familiar places and the territories passed through between them.

However, behavioral geography goes further than this. It is based on the existence of a spatial scale pertaining to each individual and on the idea that people ascribe a particular meaning to the portions of space they frequent in their daily life and across longer periods.

This has implications for how we interpret the functioning and organization of space. If space does not mean the same thing to everyone, to treat it as if it does would be a kind of violence against the individual. It would mean that central conclusions based on a singular optic would not be applicable.

In a way this subfield represents a rupture with economism and a restitution of individual values.

Yet we can hardly adopt this approach exclusively. In particular, we have to consider the economic variables of individual behavior as functions of the individual's situation in the socioeconomic scale and their position in space.

Incorporating personal conditions from a psychosocial perspective and the individual significance of space can ignore how space is defined by concrete disparities in economic possibility for individuals, in different ways and at different scales.[1] But this trend in modern geography is just beginning. As a partial approach, at least, it is full of promise, though it has not been able to demonstrate its full validity.

Perception: Subject versus Object?

Core approaches to individual perception begin inside the process of knowing, that is, in apprehending the reality contained in an object.[2] Because the principal actor in this mechanism, the subject, is at once an objective being and a microcosm, the encounter between the objectivity of a thing (or an objectified thing) and the subjectivity of its interpreter elicits multiple perceptions. The thing remains single, total, and intact, but the modalities of its perception are diverse, fragmented, and often deformed. "I am a world" (the microcosm), wrote Ludwig Wittgenstein (1969).[3]

The so-called geography of perception is limited to deepening the analysis of the perception of geographical objects. It argues that perceptions are also objective facts. It fails, however, to take two things into account. On the one hand, individual perception is not knowledge; otherwise the thing would not be objective and the theory of perception itself would be incomplete, if not useless. On the other hand, the simple apprehension of a thing, on the basis of its aspect or its external structure, shows us only the object in itself: what it *presents* but not what it *represents*. Indeed, the object is the result of parallel and concomitant determinations of structure and ideology contained within it that emerge from both the functional and the symbolic.

As Walter Kaufmann (1966) wrote, disciples of "immediate knowledge" suffer from amnesia. What they allege that they know 'immediately' is, in fact, brought into the immediate through a long historical process. What now appears self-evident was not obvious in the past. What seems simple is, in fact, the result of a long development "buried in simplicity" (23).

Livia de Oliveira (1977) is right that we should not confuse sensation and perception. We should add that we should not confuse either with the reality of the object itself that is experienced or perceived. Beyond this, she writes that "the knowledge of the physical world is as much perceptive as representative" (61).

But her work, rich in details about the biological and objective aspects of the question, neglects to mention that ideology is itself as *objective* as any other objective fact. It participates in perception and gives representation to the object observed. Experience that does not take this into account commits a cardinal sin and merely proves that objects charged with meaning transmit that meaning to their observers. The definition of an object is not limited to "receiving sensory facts and transforming them into facts of reception." Within the object perceived we need to separate the meaning attributed to it from its real meaning.

Paul Fraise (1976, 5) argues that we construct answers in a symbolic system through combining symbols as instruments of the process of knowing. When man uses symbols and signs, he does so not only on the basis of perception. Their use is not immediately oriented by things, even if everything constitutes stimuli from the environment within which man is constantly preparing himself to act.

Behavior or Praxis?

The foundations of so-called behavioral geography are broadly twofold: (a) individuals' behaviors are the outcome of individual desires and decisions; (b) individual behaviors shape space.[4]

This is an attempt to see human liberty as absolute, and not conditional.[5] The ideal of the entirely free man in a society of free men is taken as a reality. There is some confusion in behavioral geography about the scope given to individuals to choose between possible modes of action and the possibility of individuals acting arbitrarily without taking into account their income, social position, permanent or fleeting opportunities, and even place itself. The determining fact of the situation of the individual within production is ignored.[6]

There is individual praxis, and there is social praxis. But the term "organized society" supposes the precedence of collective praxes imposed by the structure of society, to which individual praxis is subordinated. Indeed space, as a locating force on both activities and men and given its characteristics and functioning—it gives to some and it denies to others— is the result of a collective praxis that reproduces social relations. When Jean-Paul Sartre (2004) refers to the counterfinality of inert material, he is talking about this: the supremacy of collective praxis (which is in fact a praxis stolen from the collective by the groups that exploit it) over individual praxes that depend upon it. To put forward alternatives that suggest absolute choice is to deny the facts. Space evolves in the movement of society as a whole. When the individual, exercising what remains of individual liberty, contributes to social movement, individual praxis can influence the movement of space. This influence, however, will always be limited by and subordinated to collective praxis. More than being contingent, behavior is a limited choice that does not change the situation of the agent, even if the action is an agent of change.[7]

Few geographers and few social scientists would have suspected that philosophical debates over perception and objectivity would interest them—certainly not as much as those about the metamorphosis of the universal into the particular, of the essence into existence, which is the basis of a phenomenology of Nature. Theoretical discussion limited itself at first to human–nature dialogue in a dualistic mode.

Two ideas persist: that perception concerns only the subject and not the object and that behavior is not related to social determinations. This is to proceed as if the individual praxis of each simply joins the individual praxis of the rest and results in total social praxis. The space of geographers ends up, therefore, reified and fragmented. Space and geography become not just ideological but inimical to the real.

7

The Triumph of Formalism and Ideology

Quantitative geography—or "the quantitative revolution"—is the central current of the New Geography. It is the product of a new era whose initial wave coincided with the end of the Second World War. It is the acme of the positivist tendency that has influenced geography since it was created as a modern discipline claiming a seat at the table of science.[1] With the technoscientific revolution, quantification was enthroned as a geographical technique, method, and even form of explanation.

Crucially, new tools made new things possible. These tools were the result of progress achieved in the exact sciences, because of both the Second World War itself and the changing needs of the new organization of the economy after it ended. Much more influential than the tools, though, were the exigencies of a new period in the history of international capitalism. This demanded the adaptation and realignment of the human sciences.

In the United States, where this realignment was most pronounced, it coincided with an epoch of veritable terror in political and cultural life: McCarthyism. This led, both inside and outside the United States, to the imposition of a series of fixed ideas, without which new economic models were not allowed to grow. They were disseminated by the new technologies of mass communication.

Geography, Planning, Utilitarianism

Imposing a new model of resource use depends essentially on two things: adoption of the idea of economic growth, and submission to a new model of consumption. Together, these two elements helped erect a new structure of production, first in the center of the system and then in the periphery. New types of consumption in underdeveloped countries helped expand new forms of production at the poles. Later, when the market was created, some production could be done in the Third World itself. As production internationalized, its autonomy in relation to consumption grew and the model was diffused more widely. Transnational companies developed in this way.

In its thousands of years of history, humanity has evolved from the coexistence of many modes of production, each adapted to the particular constellation of resources available to a collective, toward a singular mode of production divorced from local resources and directed by the needs of the center of the system.

In order for geography to contribute to this project, it had to adopt the precise numerical categories of modernization.[2] The technics of quantification became *the* method for "modern" growth.

Whether consciously or unconsciously, geography became not the guide of planning but its instrument. It justified needs defined a priori, rather than those emerging from an analysis of each given environment. The mathematical framework lent this process the guise of science.

Theses of regional inequality were accepted without deeper analysis of underlying mechanisms. Simplistic objects of calculation gave surface appearances the aura of theories or models. Reality was underestimated. It was the same story with analysis of geographical macrocephaly—the well-known indexes of *primacy*—and with the rule of *rank-size,* that became an apparently indispensable part of geographical work. It was this tendency that transformed Christaller's worthwhile theoretico-empirical exercise of *central-place* theory into an absolute rule. The clear rejection of such fixity by a reality either entirely different or clearly in transformation was dismissed in mathematical terms as a deviation to be corrected. A similar misrepresentation became possible around growth poles. Introduced into geography by Hägerstrand, the theory of the diffusion of innovations was used indiscriminately in arbitrary quantitative models, and quickly became a *marketing* tool.

These three tools have been grossly misapprehended from their creators' initial intentions. A notion of *hierarchical filtering down* has emerged:

a downward diffusion in which the growth promised but not delivered by the use of these instruments separately (development poles, the theory of central places, the diffusion of innovation) will be finally achieved by combining all three. The thesis argued that localized growth and accumulated wealth in one particular place would miraculously diffuse, after a given point, through the whole body of a nation. This miracle has never happened, in spite of the apparent proofs given by geographers using distorted calculations. In Brazil, Gauthier, a *Brazilianist* geographer, sought to "prove" that *economic distance* between regions was reducing.

If we pursue a qualitative understanding of the internal relations of any given reality, it is easy to conclude that these theories are not representative. Only a refusal to confront *quantity* with *quality* could allow such rigid models to make any sense. But a singular model, imposed from outside and with no necessary relations with the reality it is applied to, is, after all, the ideal representative form for "economic growth." In the service of this type of growth, geography must be quantitative to be any use.

The Reign of Empiricism

According to Joe Doherty (1974b, 1a),

> The adoption of an empiricist approach in geographic research was given a fillip with the "quantitative revolution" of the late 1950s and early 1960s. This trend was the outcome of a recognition on the part of some "progressive" [*sic*] geographers of the "triviality" of research in human geography which had hitherto been largely descriptive and ideographic; it represented an attempt to make geography more "scientific." The result, however, was *not* the adoption of a rigorous analytical mode of analysis, which involves (among other things) a careful consideration of assumptions and value premises (ideology), but merely the wholesale introduction of statistical procedures, leading to a form of "scientism" which is more concerned with the verification of hypotheses than with the source and nature of those hypotheses.

As a consequence, a significant part of geographical research has become fundamentally mystifying.[3]

An examination of the contemporary situation of spatial theory reveals an immense body of literature that deals with isolated or group economic activities and with the flows between them. These groupings, or nodules, are also represented by the links and flows between them: *inputs* and *outputs*. These are the facts that underpin the creation of normative and positivist spatial theories, in which man is only an abstraction, a medium or, even nonexistent.[4]

The Exclusion of Social Movement

The interests of big capital ultimately determine the groups of local activities at a given point in space. But theories of localization are based in ideas such as economies of agglomeration, external economies, economies of scale, and diseconomies of all types. All these categories are treated as if they had an absolute value that comes from their technical expression. But their meaning changes according to the socioeconomic system in place.

In fact, diseconomies do not affect companies because they are paid by the population, either directly or through the public sector. External economies today no longer need to be local. Economies of scale exist more in relation to political economy than to localization as such. Economies of agglomeration are used selectively to stop poorer companies and poorer people from accessing public goods (what is known in English as *overhead capital*).

The idea of distance, in its manifestation in current theories, is applied uniformly, as if all institutions and all companies had permanent and unimpeded access to the transport network: as if everyone were capable of using all roads and modes of transport under identical conditions. Obviously, not everyone has access to the same mobility. People are neither equally mobile nor equally immobile.

The idea of distance does not mean anything if it does not refer to the class structure and to the "value" of places, in terms of both individuals and capital.

From a social perspective, space is contoured. It is not indifferent to the inequalities that exist among institutions, companies, and people. Yet spatial theories and their derivatives—regional economics, urban economics, regional geography, urban geography, regional analysis, regional planning, urban planning, and so on—tend to ignore social structures, processes, and inequality. They end up, to put it simply, ignoring the human. These propositions cannot, therefore, become theories that transcend the ideologies foisted on people to open the way for the expansion of capital.[5]

The Ideological Vessel

For Dresch,

> In the beginning, it was as much a philosophy as a science. A philosophy that, like German historians, German geographers made to serve political ends. It was used as a medium for national and international propaganda, and as a weapon of war between States and Empires. Perhaps even more than History was. Geography still bears the marks of its youth and of the social, economic, and political conditions in which it developed. By virtue of its own methods, geography—more than any other science—bears the weight of current ideology. (1948, 88)

I agree with Dresch, but we need to add at least two things to what he wrote thirty years ago. First, geography, from its beginnings, was more of an ideology than a philosophy as such. Second, it was not only German geography that was used for political purposes. The problem was much more general. Dresch himself recognizes this, even if implicitly, in writing on the same page that geography "in the absence of methods of its own, suffered, more than any other science, from the influence of the ideologies around it." Geographical interpretation not only is "hindered by ideological currents" but itself turns into a true ideology.

Empiricism, Doherty argues, "serves an ideological purpose of the ruling class." He observes:

> empiricists do not necessarily provide apologies for the *status quo* (though they often do); thus they may not serve the *immediate* interests of the ruling class, "but those interests may be served all the more effectively by criticism which halts critical thought at superficial levels" (see Anderson, "Ideology and Geography: An Introduction," *Antipode,* pp. 1–6)—this empiricists succeed in doing by isolating the field of study, by making the abstraction

"geography" an *end* of study rather than by using it as a means by which the totality can be effectively examined. (1974b, 3)

There are some among the empiricists who, in the manner that Doherty ironically recounts, do not seem to serve the interests of the dominant classes. And there are others who, as James Anderson (1973) argues, subtly adopt a criticism of capitalist positions that in fact only barely scratches the surface.

It must also be said that so-called quantitative geography lent itself neatly to the projects of applied geographers to maintain every kind of *status quo*[6] and to the work of those who scrupulously furnished commerce with the scientific legitimacy it sought—particularly when the price was right.

I say "some" to avoid putting everyone in the same bracket. It is important to add two more categories: the substantial group for whom a "theoretical geography" of numbers and exactitude would itself guarantee high scientific quality. Then there are those who, like wind-vanes, go in whatever direction fashion dictates: they would on no account swim against the mainstream.

This last group is generally made up of opportunists. The penultimate group, though, fits the picture of numerical fanatics, drawn in another context by Cuvillier (1953, 167): people dressed in "a certain intellectual candour comparable to the statistician ... who believes it is science to count the number of widows who cross a given bridge in Paris."

For the English geographer David Slater (1975), the greatest weaknesses of what he calls contemporary "Anglo-Saxon" geography is the following:

An inverted methodology, in which the concentration of techniques imposes itself on and replace the concentration on theory; the consecration of mechanical abstractions obtained through the isolation of selected variables out of the context of socio-economic reality; the description and the measurement of forms in place of the explanation of processes; the search for a theory that is primarily derivative and a-critical; the incapacity to apprehend interconnections of vital importance between spatial structure and political economy; the imperialism of capitalist ideology that leads to forgetting that the organization of space in a determined social formation has a direct relation with the class structure present in that social forma-tion, and with its external linkages.[7]

The most serious sin of the New Geography was to restrict the horizons of the discipline and to choke off its interdisciplinarity. It is precisely by expanding geography's scope to other sciences that we can strengthen its theoretical production.

The New Geography is an involution. Based on neoclassical econom-ics, it ended up bypassing man, depersonalizing *homo sapiens* and re-placing him with *homo economicus*—a median.[8] And a median man does not exist.

The so-called New Geography also excluded the movement of society. It eliminated from its concerns the space of societies in constant move-ment. Geography turned itself into the widow of space.

Above all, New Geography killed the future. Systems analysis cannot identify tendencies. It cannot go beyond the repetitive and modelistic. Even if it is structural, it is static, and any movement refutes it. You can-not predict the year 2000 without contemporary proportions changing in your hands. You need to define new values for the variables, give them a *new quality* which will alter interpretation of their quantity.

8

The Balance of the Crisis

Geography, Widow of Space

In the first issue of the journal *Hérodote,* Yves Lacoste highlighted the crisis of contemporary geography. He listed the fundamentals that would allow him to speak of a geography *of* the crisis (1976, 8–69): "It is as if everything blocks the path to producing a concept of space. Perhaps this is precisely because of the gravity of the political conjuncture, and a collective and unconscious refusal to reflect on it" (50). Other geographers have also warned of this crisis: David Harvey, German Wettstein, Richard Peet, David Slater, and many others. More recently, as new tendencies have begun to emerge, Gilles Sautter (1974) has asked whether this crisis can lead to renovation. Yves Lacoste is somewhat optimismic. This is not to say that the traditions of the past have abdicated their place and the crisis is being overcome. It is still the case, as Paul Claval wrote in 1964 in his *Essai sur évolution de la géographie humaine,* that "there is a malaise in contemporary geography."

The Reproduction of Knowledge

Geography's development has been so slow and underwhelming that at times it seems like involution, not evolution. One of the main reasons is the sheer weight that old ideas have within the discipline.

Attachment to timeworn notions seems to be an incurable sickness. Geographers are well known for their penchant for clinging to a problem or a theme for years and years. Looking at a geographical bibliography for the twentieth century, you see how rare, imaginative themes are swamped by a wave of repetitions.[1] A neutral observer could justly accuse us of a serious and repeated failure of the imagination. Our defect is what David Bohm (1965, 9–10) excoriates as hindering the development of science as a whole: we see old ways of thinking as inevitable. Rather than pursuing new knowledge, we prefer to amuse ourselves with the reproduction of old knowledge. This is made possible by a kind of co-optation that, though manifesting in various ways, ends up producing a widespread canonization of models. Jacques Lévy (1975, 64) puts it well: "The co-optation of ideas," he writes, "plays out the co-optation of people. Well-established geographers struggle to maintain the dominant ideas, regardless of how indefensible they are." Leading professors (those the French call, blithely, "the bosses") manage access to the teaching profession, act as gatekeepers of research, and control the organization and diffusion of publications. This gives them an unalloyed prestige that impedes genuine critique of their ideas. Indeed, it encourages their ideas to be reproduced, however dated they might be.[2]

Writing recently about the conditions of the evolution of French geography, Claval outlined how the taste for bureaucracy was acquired and became institutionalized. He described how it encouraged hierarchical relationships that distorted both the objectives and the results of research. Research, he argued (1975, 262), "became more conservative than before," and its timidity "was all the more lamentable because the traditional structures of French universities were profoundly liberal, . . . encouraged boldness and provided the conditions for individual ingenuity." He continued, "anyone who proposed new ideas found their hypotheses under attack and their work viewed with suspicion. Ultimately, though, their ideas were plagiarised by those who had previously condemned them. This kind of late adoption nevertheless guarantees its subjects a place in the geographical community" (262).

Addiction to hierarchy is most obvious when a great thinker's ideas are misrepresented or misconstrued by people they cannot directly influence. Christaller's followers in the United States, for instance, misrepresented his work. Ratzel's work was praised by Vidal de La Blache but later repudiated by La Blache's pupils in their proclivity for national schools.

The historical development of geography, then, has always constituted more of an encumbrance than an aid.

Geography and the Imperial Project

The tendencies outlined here became exponentially worse after 1945, when geography was put ever more clearly at the service of an imperial project. The center of dispersion of geographical ideas began to shift from Europe to the United States. For those who like to mark transitions with dates, we could identify the International Congress of Geography in Washington, D.C.

The postwar period marked the beginning of American economic supremacy. The production of ideas was concentrated in the United States, and American English became hegemonic. Overwhelmingly, the ideas produced were authoritarian, destined to serve commercial and political projects. This guaranteed their partiality.

At an earlier point in history, believing in evolution was enough for scholars to be expelled from universities.[3] Later, McCarthyism infiltrated all corners of American life, not least universities. Nowadays educational institutions in the United States congratulate themselves on the liberty with which their professors and students work. Yet certain interpretations of reality—principally dialectical interpretations—are practically prohibited, and many of those who insist on dealing with these subjects are left unemployed.

It is very difficult to reconcile a systematically discriminatory attitude with the flourishing of the scientific spirit. These tendencies emerge strongly in geography: our far-reaching discipline has become a valuable tool in the technocratic processes through which rich countries have ever more effectively exerted their dominance over the rest.

In the postwar period, we can say of the underdeveloped countries the same as what Chastaign said of France in relation to Germany after the Second World War: they have been occupied by North American thought. In geography, Latin America is a good example, but not an isolated one. In fact, to speak of "North American" ideas about the problems of the Third World is to generalize too far, because some North American geographers and social scientists work with sincerity and competence. But there is an official geography that controls both the means of the diffusion of knowledge (editors of magazines and books, international and national associations, congresses, and so on) and the means of the production of knowledge (such as grants research resources).

Abstract Empiricism

For modern positivists, who banish all concern with history, things have a definitive value "in and of themselves." The process of their formation is of no interest. They are more concerned to demonstrate how things are than to explain how things are. The preoccupation with measurement imposes itself on the search for the real causes of things (Kopnin 1966, 1969, 69).

The great poverty of empiricism is that it hides the relations between people and substitutes them for relations between objects, including geographical objects. But things by themselves don't possess relations. All projects that fail to take this into account replace a true meaning with a false one. This is a form of abstraction without basis in the real. It is an abstraction with a false origin because it is grounded not in the essence of things but in their appearance. Empirical abstraction in geography can take on an open, brutal form, or a more subtle and sophisticated form. Of the second, a good example is the center-periphery model critiqued by, among others, myself (1975) and Daniel McCall (1962).[4]

In their classic article "Capital-Labor Substitution and Economic Efficiency," Kenneth Arrow, Hollis B. Chenery, Bagicha S. Minhas, and Robert M. Solow (1961, 225) give a clear example of the use of empirical abstraction in theorization: "In many branches of economic theory, it is necessary to make some assumption about the extent to which capital and labour are substitutable for each other." They add: "Given the lack of empirical generalizations of the phenomenon, theorists choose simple hypotheses that become generally accepted by being frequently repeated."

Norton Ginsburg (1973, 2) has strongly criticized this tendency as elaborate theories based on pedantic abstractions from reality. "It is so abstracted from reality that we hardly recognize reality when we see it. If one were to seek in geographic theory the essence of American society as it exists today, one would be hard-pressed to deny that it exists in order to consume, not the other way around." Joël Pailhé, Christian Grataloup, and Jacques Lévy (1977) are right when they critique a geography "for which space, like time, is not an objective fact, does not have a real existence, but is located in our way of understanding things." Therein is the idealist philosophy that has lain deep within geography since the end of the nineteenth century. Dressed up as concrete objectivity, it triumphed at the beginning of the twentieth century. The theories of the diffusion of innovations, of central places and of growth poles, the principle of "least resistance," the law of spatial gravity, and so many other weapons in the arsenal of the sciences of space were donated to the armory of planning. These are nothing less than explicit or implicit forms of philosophical idealism, if not of abstract empiricism.

Geographical space, then, is studied as if it were not the result of a process in which man, production, and time played essential roles. The space of man must be debased in order to give the impression that, in the act of production, men are confronted with a dehumanized, parceled-up, and reified space. Real space is substituted for ideological space, under which false theories and planning can hide.[5]

So-called quantitative geography demarcates the maximum extent of this despatialization of space, reduced to a net of coordinates with no relation to the real. It becomes a "*computer-taxonomic exercise*"[6] (Brookfield 1975, 107), a dehistoricization, a collection of mathematical formulas from which history—or, indeed, man—has been systematically erased. It marks the deformation into sophistry of an old trope of geography's positivism: the notion of the region. This was a preliminary, but fundamental, step.

To study the region as an autonomous entity is to make it the object of a defective dialectic, in which man is separated from capital and the environment is separated from society. This replaces a dialectic in which men are studied on the basis of the internal dialectic of society as a whole. A distorting emphasis on landscape is of a similar order. Recently Claval (1974a, 42–43) has noted that the most important aspects of the study of landscapes have been lost and the least important ones put in the foreground. Sorre (1957, 31) insists that it is a question of the means of explanation: "behind the concrete traces of the landscape, our analysis reveals a whole network of causal relations." It would otherwise be devoid of "scientific value."

As geography itself falls further behind, the city has emerged as a terrain of encounter for sociologists, economists, anthropologists, ethnologists, politicians, and historians. Regional analysis gives rise to a proliferation of specialized disciplines, not least to answer the needs of planning. The task of formulating general principles has been gradually abandoned by geographers. Thus geography has reduced its own field of action, interest, and significance, while the disciplines that surround it little by little fill the void. In general, however, they cannot truly address the question of space.

From Imperialism to the Loss of the Object

In 1836, Ritter was already complaining of how little geographers had achieved in his field of interest.[7] In that period a truly scientific geography still did not exist, thanks to the lack of two essential conditions: the world itself was relatively little known, and the social sciences were yet to be constituted. By 1921 the Belgian geographer Paul Michotte was already lamenting that geography was in retreat.

This reduction of the field took place in parallel with the emergence of the tendency for geographers to think they could address anything.

Omer Tulippe was right when he called out geography's imperial vocation. He wrote in 1945: "after this initial confusion, the field of geography gradually rebuilt itself. However, in the course of this reconstruction, the limits of the field were extended into the distance. This is where the overly ambitious, encyclopaedic attitude of science comes from; an attitude that also afflicts geography" (74). For Tulippe, "this caprice is a basic error, a sin of youth, which happens to other sciences too, but which geography is only slowly untangling itself from" (76). Whether thanks to the incursions of other disciplines into its field of study or to its own incapacity to define an object of its own, geography faced the problem, in losing its breadth, it lost almost everything.

S. H. Franklin in 1973 declared that he was "always apprehensive that the next history of geography that I read will in fact be an obituary. Repeatedly geographers discard, sometimes they murder, the vital topics" (207).[8] His concern was heartfelt.

Space Pulverized

According to Michotte (1921), "the progressive division of labour in science led to a progressive and ever clearer specialization, destined, in the case of geography, to strangle its object."[9]

To use Dresch's words (1948, 91), we end up with a geography "cut up into little pieces."[10]

Since the creation of geography as a discipline claiming scientific status during the first half of the twentieth century, two essential tendencies have run through it. On the one hand, some authors have worked to secure geography's place as a science through discovering laws and general principles, defining their field, classifying facts, and establishing a hierarchy of values. On the other hand, there are those who have variously sought to make geography into a body of immediately applicable knowledges without worrying what the actual or potential demands of the users of such knowledge might be. The former is a speculative approach, and the second is pragmatist. There has been excellent work in the United States— that of Richard Hartshorne, Carl Sauer, and Fred Schaeffer stands out, and there are others—but we can nevertheless argue that the former category is largely European. This explains why European geographers tend to see space as a unity, even if they rarely transcribe their intentions into theory and method. American geography, fed by pragmatism, takes isolated, singular aspects of reality as its object, depending on the demands of the client.[11] This ends up decimating both the object of the discipline and the discipline itself. The proliferation of themes to study distances it ever further from the construction of a synthesis. As geography becomes more and more utilitarian, it becomes less and less explanatory.

Geography, Widow of Space

Geographers everywhere have been, at least to some extent, silent about space. Sometimes they have also been silent about the innovative work of other geographers and students of space.

Geography is the widow of space (Santos 1976). The fundamentals of its teaching and research are orthodox historiography, "natural" nature, and neoclassical economy. Static, absent, and ideological, all three have been placeholders for the real space of societies in constant processes of becoming.

This is why so many geographers have talked so much about geography—a word increasingly emptied of content—and so little about space as the object and content of the discipline of geography. Defining space as an object has therefore become difficult, and defining geography has become impossible.

Detemporalizing and dehumanizing space, geography ended up turning its back on its object and becoming "the widow of space." Partly this is because of a drain of energy and talent into the search for immediate solutions to apparently immediate problems, and particular responses to apparently particular problems. We end up, therefore, with a great multiplicity of geographies. Lots of geographies, no geography.

Put at the service of things and not of people, the search for identity and scientific legitimacy stopped being a driving force and became a hindrance to a project oriented more and more toward commerce, not social interest.[12] Self-reflection became unnecessary.

In fact, as Sorre wrote (1957, 51), geographers lost the taste for debate that characterized their heroic period.[13] Substantive debate was exchanged for tiffs over form. An interminable quarrel over vocabulary replaced the search for fundamental mechanisms.

And so we witnessed the deterioration of geography into an incoherent and formless compendium, a *puzzle* to suit the interest of the consumer. This is depressing, because even if other social scientists serve vested interests with blind loyalty, they at least do so with a modicum of general theory.

Lessons and Promises from the Crisis

It is no wonder, then, that the general crisis of the social sciences landed with vigor in geography. This crisis should not, as it often is, be prudishly hidden under the pretext that this will protect the discipline from lethal criticisms. In fact, these have been coming from all sides for a long time. They have recently become more vigorous, with the expanding debate over disciplinary objects seen as imperative within the contemporary transition to a world society.

Instead of denying the crisis, we should expose it. A state of crisis is not a testament to weakness. Crisis proves that the old is fragile and is protecting itself from the new that threatens to replace it. Those who are trying to find new ways forward, including scientists, must be constantly on the alert.

For geography, the crisis deepened because as mistakes accumulated, errors crystallized. Each step forward turned out to be a huge step back. This is the very definition of geography since its foundation, and above all since the Second World War: a set of statements that, instead of helping to discover the real, helped to obscure it. This set of statements was hiding within a scientific rhetoric. In a period when science was seen as the study of phenomena and appearances, this rhetoric—necessary to the smooth expansion of the capitalist system—impinged on the analysis of reality itself. A scientism unconcerned with the essence of things was both the stimulus for and the justification of an empirical geography. It was merely a screen for the crisis.

Compromised by vested interests, the tendency of science to obscure an accurate image of reality amounted to concealing the evolution of human history. But history cannot be ignored in the end. It overthrows the myths that are created to shroud it. This is one of those historical periods when a fundamental change in the nature of space demands a fundamental change in the definition of space. Explanations elaborated outside reality become innocuous and insist on being replaced. Such a moment is also full of attempts to replace one lie with another. In the face of the new reality emerging in front of our eyes, we make a desperate effort to find a false theory. To avoid falling into the errors of the past, we must insist on the existence of the crisis. But we must do so with a critical mind. We must examine not its appearance but its foundations.

PART II

Geography, Society, Space

9

A New Interdisciplinarity

Ever since geography began to identify its unique scientific qualities, geographers have believed it to be, above all, a science of synthesis, singularly capable of interpreting earthly phenomena using tools from the natural, exact, human, and social sciences.

Some geographers would go even further to affirm—an oft-repeated trope—that geography is the *only* science capable of such a synthesis. This claims that the explanation of geographical phenomena, more than any other discipline, requires the contribution of many sciences.[1] The sciences summoned to help geography have been referred to as "crutches" or even, apparently without irony, as "auxiliary sciences." Here, geography is the conductor, and other scientists are the musicians. This outlandish self-aggrandizement and frankly insufferable pretension ignores the fact that geography has never in fact developed the instruments necessary to realize such a synthesis.[2] It is ridiculous to argue that one discipline is the best. All sciences are synthetic, or they are not sciences. As Harold Brookfield wrote in 1973: "We are not better than anyone. We are equal to all the other sciences. The geographer is no more nor less capable of elaborating a synthesis by virtue of being a geographer."[3] The capacity for synthesis is not reserved for any particular specialist. It emerges as a result of an intellectual preparation that goes beyond one's own specialism to embrace the universe of things and to understand each thing as a universe.

The Isolation of Geography

Geography faces a paradox and an irony. If it is a synthetic science, it is one that in its day-to-day practice has little to do with other disciplines. Its isolationism is partly responsible for the difficulties it faces in evolving.[4]

Edward A. Ackerman (1963, 431–32) observes this failure, arguing that only a few geographers have made more than banal generalizations about the universality of the scientific method.[5] Excessive generalization has paralyzed geography's development and hindered the search for a true geographical theory and methodology.

Perhaps it is fitting to turn to William Bunge (1966, 142) here: "To hope for an intuitionist theory of intuition is as ingenuous as to hope for a mystical theory of mysticism or a schizophrenic theory of schizophrenia." If we begin only from geography we will arrive at an intuitionist, mystical, and schizophrenic geography. The old idea of national schools of geography—French, English, American, and Swiss—is part of this schizophrenic mysticism. But the values of geographical research have been reproduced in many countries. Even if conditions in different countries—internally and in relation to the rest of the world—have lent geographies particular textures, they have always been deeply interrelated.

Defending national schools is about cross-border competitiveness. Indeed, so-called national schools of geography work more effectively abroad than at home. They propagate cultural imperialism and, through local intellectuals, insidiously spread interpretations alien to local realities.

National schools contribute to political, economic, and cultural colonialism. It is relatively easy to work out whether any given African geography has come from a country colonized by France or by England. Latin American geography, long defined by French influence, has more recently obeyed North American diktats.

But the history of geography is full of ironies. Anything can happen. The export of knowledge-making that represents the internal and external interests of an exporting country can end up reverberating within it. Teaching and research are intimately linked and also connected with dominant national political economic interests. This helps create an isolationism intensified by linguistic barriers and the deepening of the struggle for hegemony between the rich states.[6]

Evidently, this phenomenon is not the same everywhere. When Joseph Schumpeter (1964), for instance, noted that French economics was largely self-sufficient, he added that this was different in Germany and had its roots in the early development of sociology in France and its influence on economics.

The Advantages of Interdisciplinarity

Geography, more than other disciplines, suffers from a weak kind of interdisciplinarity. This is linked to the diverse and multiple nature of the phenomena that geographers work with and to the university training they undertake.[7]

At the beginning of the nineteenth century, Carl Ritter identified the need for interpenetration between scientific disciplines.[8] He argued that this could take many forms. He himself was educated in philosophy and brought up on the teachings of Hegel. He thought of interdisciplinarity as necessary for the universal aspirations of his age and an output of the growth of scientific knowledge during the first technoscientific revolution.

The coherence of the drive toward interdisciplinarity among geographers is yet to be properly analyzed. This is a difficult task because they are all, or nearly all, convinced that they already work in an interdisciplinary way.[9] This isn't really true, so geography does not benefit from interdisciplinary enrichment.

The English philosopher Alfred North Whitehead (1938, 136) notes that explanations for phenomena from one science are often shared in another.[10] In other words, we cannot reach a valid explanation if we merely remain confined to sociology to explain a social fact, to economics to understand economic phenomena, or to geography to interpret geographical realities. There is no reason to fear the invasion of other specialists. In relation to the economic, social, and political dimensions of the process of development, Ernesto Cohen (1973, 4) wrote that "concepts endogenous to one dimension are exogenous to another." Indeed, when an exogenous fact is incorporated into the interpretation of some aspect of reality, it immediately becomes endogenous to that explanation. Schumpeter testifies to the importance of the nonprofessional in the progress of any science (1943, 1970, 45), for all progress owes thanks to the trespasses of specialists from other disciplines. This is what Jean Chesneaux (1976, 164) calls "stealing privileges from the professionals."[11]

The great French historian Marc Bloch (1974, 166) wrote that "the sociologist Durkheim and the geographer Vidal de La Blache left a mark on historical studies at the beginning of the 20th century far deeper than any historian." Maximilien Sorre and Pierre George, two geographers, did the same for sociology.

Geography itself can contribute to the conceptual evolution of other disciplines, such as economics. This has become even clearer since neo-classical economics has become scholastically, instrumentally, and polit-ically essential to the diffusion of capitalism. As neoclassical economics is, by definition, an abstraction from man and the geographical environ-ment, geographical studies have significantly helped hone many economic concepts.

In the early 1970s I proposed the concept of a double circuit of the economy of underdeveloped countries (Santos 1970, 1972, 1975a). I arrived at this conceptualization because I was unable to find a dynamic interpre-tation of geographical space in official economic categories. André Marchal has argued that "economic laws are nothing more than the reflection of the behaviour of men. And this behaviour varies according to periods and places." In this case, what is true for economics is true for geography.

The principle of interdisciplinarity is common to all sciences. As Jacques Boudeville wrote, "all sciences develop on the boundaries of other dis-ciplines, and merge with them into a form of philosophy. Geography, sociology and economics are complementary interpretations of human reality."[12]

Geography and Interdisciplinarity

The search for interdisciplinarity proposed long ago by Ritter has led geographers to reach some particular conclusions. One was the corona-tion of "special geographies," a formula adopted by Brunhes and Vallaux, both critiqued by Sorre. For Vallaux, the problem of formulating the sci-entific basis of geography did not rely directly on so-called auxiliary dis-ciplines but drew on particular geographies that emerged from them. S. V. Kalesnik, more recently, has reiterated Sorre's position. He objected (1957, 36) that "each of the basic phenomena that general geography is concerned with belong, without any doubt, to autonomous disciplines, but their field and their research processes are different." Therefore, he wrote, "I do not accept the idea that [general geography] is the negation of the unity of human nature." "In spite of what Brunhes and Vallaux said," Sorre argued, "there is no special geography, no special geographical problems, but only chapters of a human geography whose unity cannot be broken because the individual human is, in every one of his actions, a total human."

According to Sorre, Jules Sion had arrived at the same conclusion but had unfortunately been misrepresented by other geographers.[13] Other attempts at interdisciplinarity failed because geographers' knowledge of related disciplines was thin and incorporated without adequate conceptual apparatus.[14]

As in other disciplines, the holy grail was never found, because of the confusion between multidisciplinarity and interdisciplinarity. Multidisciplinarity suggests that a phenomenon be studied through a multilateral collaboration between disciples. This is not the same as an integration of these disciplines, which is realized only through interdisciplinarity: an imbrication between diverse disciplines pertaining to a collective object of study.

The confusion between multidisciplinarity and interdisciplinarity has been abetted by the conditions of organization and operation of universities, in particular in the United States. Consciously or not, there has been an urge to present little bits of realities pasted together into a collage as if they were a unified reality, or, indeed, reality itself. But interdisciplinarity cannot function like a patchwork quilt. Many North American universities follow a commercial model that overdetermines their pedagogical objectives. "Interdisciplinary" courses are a formula to increase the profitability of the university enterprise. Lots of students can be packed into one room, in front of one professor. Revenues go up without costs increasing. As students operate at different levels and come from different areas (medicine, engineering, dance, sociology, economics, fine arts, and so on), the only possible form of interdisciplinarity is an epistemological one. But the teaching is, above all, factual and empirical. Mercantile interdisciplinarity, far from aiding the progress of science, contributes to its regression. This commercial model of universities in the developed world can be transplanted to countries with extremely diverse real-world conditions.

This wider problem is compromising scientific development in general, but geography, with its specific characteristics, is particularly badly affected.

The Stages of Interdisciplinarity Applied to Geography

We can define three stages—all abortive—in the history of interdisciplinarity in geography, and a fourth which is currently emerging and which I want to help to delineate.

First, there is classical interdisciplinarity, based on bilateral relations between geography and history. History and geography were long thought of as identical twins. At the beginning of the nineteenth century, Kant (1802, vol. 1, 6–8) wrote that "history concerned itself with the description of happenings according to time, and geography with the same object, but according to space." Thus, history would be different to geography only thanks to differences between time and space. History would recount events as they succeeded in time; geography would concern itself with how they took place in space. This Kantian position held sway for a long time and is still influential, as if, after Einstein, it were still possible to maintain a Newtonian separation between time and space.[15] This has led to serious flaws in the theoretical and methodological progression of geography.

The notion of a history that organizes phenomena in time and a geography that organizes them in space is the inheritance of Kant, perfected by Alfred Hettner. Countless contemporary geographers leave this idea almost intact. It is responsible for a grave error of method: geography should concern itself with researching how time becomes space and how past time and present time each has a specific role in the functioning of present space. "Geography should take into account the social formations within which it places the differentiation of social space. History should not be imagined as a factor that explains geography (i.e., historicism) but, on the contrary, geography is that which is considered historically" (Pailé, Grataloup, and Lévy 1977, 49).

For H. C. Darby (1953), it is impossible to trace a line between geography and history, because "the process of becoming is one process."

The problem is often badly formulated. It is taken as if it were a question of the relations between history and geography. This is a false start, with false premises. When writing history or geography, each of these terms comes loaded with principles and preconceptions that, if accepted, remorselessly find their way into whatever follows.

It is better to think in terms of space and time. These ideas are not free of their own difficulties. Indeed, they can be just as fraught as the lexicon of history and geography, because the debate over the meaning of time and space is as old as philosophy itself.[16]

A second stage of interdisciplinarity in geography is noteworthy more for a negative than for a positive reason: the refusal of geographers to refine concepts originating in other disciplines. This phase coincided with the crucial moment in which the founders of modern geography came to adopt the principle that geography should become an autonomous science.

Vidal de La Blache, founder of modern French human geography, utterly rejected Durkheim's proposal to include geography in a basic classification of the social sciences. The idea of social morphology—a sociological discipline dealing with the modalities of the transformation of society in geographical space—was aberrant to Vidal. It provoked a debate whose consequence was a prolonged separation between geography and sociology. The result was an impoverishment of geography and the birth of parallel disciplines whose concerns could have lain within geography if disciplinary disputes had taken another direction.

More recently—and this is the third stage—the necessary interdisciplinarity of geography has acquired a new dimension thanks to two important historical advances: the first at the end of the nineteenth century, and the second, more recent, and more brutal, after the end of the Second World War. These developments have deepened the hegemony of technology and broadened the field of the social sciences.

The notion of interdisciplinarity evolved with scientific and economic progress. New realities, requiring new explanations, demand the appearance of new scientific disciplines. This leads to the death of the classic form of interdisciplinarity and its replacement by another. That which yesterday could still be considered a reasonable interdisciplinary focus can no longer be so today. It has become necessary to refuse partial contributions that used to be useful but no longer reflect reality. Under new conditions, the possibility of helping other disciplines develop from the outside, using new materials, increases. Advancing a given science is no longer the sole privilege of its specialists, but it is crucial that scientists be predisposed to interdisciplinarity and equipped with the critical faculties guaranteed only by a coherent philosophy.

This idea should be at the front of our minds when we work with historical phenomena like space.[17] The list of sciences related to (human) geography is familiar: history, sociology, and economics. But this list should be greatly expanded to include technology studies (the science of the productive forces), political science, urban studies, management studies, semiology, epistemology, international business studies, the history and philosophy of science, and even logic and dialectics.

Technology emerges as a science the moment the productive process requires a predictive capacity—that is, the need to say, in advance, what is going to be produced, how, why, and with what results.

Technology becomes history through the intermediation of technics. Technics are the modifying intermediation between nature and human groups.

Each technical evolution corresponds to a new form of organizing space. We cannot reach an understanding of space without a precise comprehension of the meanings of the instruments of labor: houses, streets, factories, dams, and so on.

The transformation of technology into technics is subordinate to ideological, economic, and political givens. We need to intervene, therefore, in the teaching of these sciences. Today, ideology's role in the interpretation of space has deepened because the objects to be planned and constructed appear to have a meaning that they do not, in fact, possess. This apparent meaning often emerges from a concern with international interests. So we need to understand international relations. And in order to separate the meanings given to objects from their real value, we need to understand semiology.

Equally, the social elements of collective activity and the construction of space are addressed by anthropology and sociology.

The particular characteristics of each population influence spatial evolution. And so we need to understand demography.

This list will also need to be extended to include the principles of physics. Jean Gottman (1947, 5) has urged us not to forget the essential difference between inert material and human material, which is living and active.[18] This is less a positivist search for an analogy between physical phenomena and social phenomena than a deployment of the philosophy of physics. This tendency cannot be exaggerated. Bertrand Russell

(1958, 134) once wrote that "broadly speaking, traditional physics has collapsed into two portions, truisms and geography." Still, if the sciences explain the visible and the phenomenological, only philosophy expresses the real.[19]

The differential development of the sciences has magnified the task of establishing a valid interdisciplinarity and generated a fear of confronting it.

Ely Devons and Max Gluckman even argue that the intrusion of a scientist into another discipline is dangerous ... except for geniuses.[20] This is obviously an exaggeration. Harvey (1972), one of the few geographers to venture into the intricate field of the epistemology of geography, argues that we need to understand psychology, economics, sociology, physics, chemistry, and biology. He fears that (1972, 41) "the sheer necessity for specialisation is bound to force many of us to concentrate on just one aspect of our grand theoretical design." This would lead to exactly the opposite result from the one we want because, instead of securing an interdisciplinarity through which we could understand multiple aspects of the same object, we would deploy a hasty interdisciplinarity, a false specialization with failings akin to those of mechanistic analogies.

At the other extreme, Anderson critiques how many disciplines are concerned with the same object (1973), lamenting how many social disciplines—geography, anthropology, economics, political science, demography, administration, psychology, sociology—are turning their attention to the study of the city but that all of them are more or less ignoring the others and "working with their own methods and creating their own methodology."

Evidently, this does not help us to construct new scientific understandings.

The line between the use of a discovery obtained in one field of knowledge and a full and deep understanding of that field is well placed by Whitehead when he refers to the enormous contribution of Einstein to the development of the sciences as a whole. In a speech given to a group of chemists he said: "I remember that I am lecturing to the members of a chemical society who are not for the most part versed in advanced mathematics. The first point that I would urge upon you is that what immediately concerns you is not so much the detailed deductions of the new theory as this general change in the background of scientific conceptions which will follow from its acceptance" (1964, 164).

H. Reichenbach ([1920] 1965, 1) was right when he argued that Einstein's theory of relativity had a positive effect on the fundamental principles of epistemology.[21]

The Need to Define the Object of Geography

The fact that doing legitimate interdisciplinary work is so hard has bred a consensus in favor of cooperation. Specialists from various areas can be brought together, each bringing his own methodological toolbox, and together they can enable geography to be truly interdisciplinary. This solution, the argument goes, could work for other disciplines and they too would become interdisciplinary. Indeed, if we were to accept Julian Huxley's advice (1963, 8), each researcher, with his own point of view, methodology, and technical lexicon, would abandon work on particular problems to build networks of cooperative research with common methods and terminology. This would, eventually, come together in a global process of research (see Harvey 1969, 123).

The Belgian geographer P. Michotte wrote in 1921 (see Fisher et al. 1969, 185) "that it seemed very doubtful . . . that one human brain could hold on to a total vision of the world and its laws." If it could, it would be "in a very superficial form." Future efforts to synthesize reality would be less a synthesis in the etymological and literal sense of the word and more a kind of encyclopedic adventure.

At first sight, these arguments are tempting. They are logical and appealing. They propose that, rather than understanding one aspect of total reality through a particular discipline, we need an understanding of the whole through the whole.

Understanding the totality is fundamental if we are to apprehend the ontological and epistemological place of its different parts. But the knowledge of the parts—their functioning, internal structure, laws, relative autonomy, and evolution—is a fundamental instrument for the knowledge of the totality.

Interdisciplinarity that does not take the multiplicity of how reality appears to us into account is risky. An ungainly and compromised theoretical construction cannot correctly define the parts of the totality and can further complicate the problem of how to define total reality.

Indeed, such interdisciplinarity presupposes that the object of geographical inquiry has been ascertained and its basic categories identified. Of course, the meanings of categories change across history, but they remain a permanent basis for, and a guide to, theorization—in our case, the theorization of the production of space. If we do not recognize this, we fall into what Harvey (1969, 122) has called "*intellectual dandyism*": the search for superficial explanations based on a fragile understanding of adjacent disciplines. For geography to be truly effective, it must begin not from the interpretative elements of other disciplines but from the object of our own discipline, space, and how it appears as a historical product.

10

An Attempt to Define Space

Durkheim (1962, 18) argued that the sociologists of his time worked with concepts, rather than with things. If we are more concerned with geography as a formal science than with its real object, space, we risk the same charge.

To Define Geography, or to Define Space?

The challenge is to define the object of each discipline within the totality of knowledge. In geography, this theoretical and methodological exercise raises a number of risks, most seriously that of confusing science itself with the object of science.

In 1925 Martonne wrote of the ties between our discipline and other branches of knowledge. However, he was referring to the relations between geography and other sciences, not the relations between the object of geography—space—and other tangible or intangible elements of social reality. Martonne's starting point necessarily leads to an interpretative fallacy. Particular sciences produce holistic knowledge of particular aspects of reality. The evolution of scientific thought enables more precise knowledge of these diverse aspects and, in parallel, a better understanding of reality as a whole. This operation is multiplicatory. Each new synthesis advances analysis of each particular aspect, and vice versa. Unfortunately, however, of all the social disciplines, geography has been the slowest to define its own object and has, therefore, completely neglected this synthetic approach.

Hartshorne was one of the most influential geographers in the United States, not least because of his theoretical dynamism. He asserted (1939, 374) that "geography is to be defined essentially as a point of view, a method of study," rather than in terms of its object. The French geographer Maurice Le Lannou went further, affirming categorically that geography was only "a point of view."

This defining (or not) of geography's field of interest has influenced the way our discipline is seen by other specialists. In 1969, the geologist P. Rat wrote: "it can be said that there are no geographical facts, but a geographical way of considering each set of facts."[1] Others, like C. R. Dryer, think in terms of the distribution of things over the face of the earth, but they also put things in their context (see Freeman 1961, 70).

The many definitions of geography are, therefore, unhelpful for its development. Lukermann (1964), for instance, thinks that neither content nor method are important and that geography is defined by the questions that it poses (cited by Minshull 1970, 11). For the English geographer Clements Markham, these questions would be: "Where is it? What is it? How is it? When was it?" (1905, 58).

Reproducing lists of definitions of geography is always tiresome, and sometimes counterproductive.[2] Even if a science is defined by its object, it is not always the case that the definition of the discipline takes this into account. Geography's concern with its explicit object—social space—was always secondary. Yet there can be no scientific progress without reflecting on how different aspects of reality are studied.[3]

The ultimate concern of each and every branch of human knowledge is *total* society. Each particular science concerns itself with one of its aspects. The fact that society is global enshrines the principle of the unity of science. Total social reality presents itself to each of us in any given time and place only through just one, or some, of its aspects. Hence the existence of particular disciplines. This is not to gainsay the principle of the unity of science but to set up another fundamental principle: the division of scientific labor.

According to Fraise,

Each science . . . corresponds to a level of the organization of nature. Between these levels there are both continuities and discontinuities. Each science is, by its own divisions, reducible to the analytical capacities of an inferior level. In terms of specific organization, though, each remains irreducible. Psychology cannot be reduced to biology, nor this to chemistry, even if there

is a place for biochemistry and psychobiology. Each level does not correspond to a new ontological entity but to an organization whose properties
reveal themselves through the behaviours that they determine. (1976, 11)

Georg Simmel (1894, 1898), among others, tried hard to delineate an
object for sociology. He endowed the discipline with qualities from other
humanities. Meanwhile, concern with principles and classifications has
meant that the actual content of a new science of geography has been
obscured.

As Durkheim (1900) wrote, it's not a question of establishing precise
limits to science. "The section of reality which [a science] proposes to
study is never precisely separated from others. Everything in nature is connected, so we cannot impose strict boundaries between different sciences,
which are in fact continuous with one another."[4] But if we cannot recognize what the domain of the science is, we risk—as Durkheim warned for
sociology—merely endlessly extending geography's remit (1900, 1953, 179).

The Problem of Autonomy and Analytical Categories

The relative autonomy of each discipline can be found only within the
system of sciences whose coherence is given by the unity of the object of
study itself: total society. Yet the coherence of each discipline requires a
system particular to it, formulated through previous knowledge of a part
of social reality understood as a lesser totality. The parcel or aspect of
social life considered in this way comes to be the object of each particular discipline. Without this approach, we could not construct knowledge
about what we study and attempt to understand.

This reemphasizes the need to identify the true object of geographical study. However, identifying this object is irrelevant if we cannot define
its fundamental categories. From a purely nominal perspective, categories
change their meaning across history, but they also constitute a permanent basis for and a guide to theorization. If this indispensable exercise is
to be effective, we must concentrate our concerns around the category—
space—as it presents itself as a historical product. The primary facts of
interest are about the genesis, functioning, and evolution of space.[5]

Our interpretation of space and its genesis, functioning, and evolution
depends on a prior correct definition of its analytical categories. This

enables an analytical disaggregation of the whole, and then its synthetic reconstruction, and helps identify which other scientific disciplines might be relevant. What type of collaboration can they offer us? What use can we make of their teachings? Because it is neither all sciences, nor all of any science, that enable interdisciplinarity.

As reality is a totality in permanent movement and transformation, the disciplines involved in interdisciplinary efforts are constantly changing. This is due to objective factors as much as to the judgment of the researcher. There are many reasons why sciences collaborate: (a) scientific progress, responsible for the creation of new disciplines and the evolution of existing ones; (b) the philosophical and ideological position of the researcher and his vision of the object of their discipline; (c) the historical context, which can prompt researchers to pay more attention to conjunctural characteristics—which can deform the image of reality—over structural characteristics.

It is not difficult, therefore, to establish a direct relationship between interdisciplinarity and the epistemology of each science. Epistemology is a philosophical reflection particular to each field of knowledge. However, it is not immutable. It functions as a kind of policing, such that ingredients of multiple origins do not confuse the specialist but allow them to uphold the scope of their own research. This does not mean that the object of each particular discipline should be rigidly determined, incapable of evolution and change. The great merit of high-quality interdisciplinarity is that, while it disciplines work within a particular science, its productive contact with other parts of knowledge opens new pathways.

Whitehead expresses this in magisterial terms: "Even within the circle of the special science we may find diversities of functioning not to be explained in terms of that science. But these diversities can be explained when we consider the variety of wider relationships of the pattern in question" (1938, 136–37).

The Scientific Object and Theorization

To define the object of a science is to construct a system to identify the analytical categories that reproduce in the *ideational* field the totality of processes these categories produce in *reality*.

Each science's categories of analysis must align with the categories of reality. This is to approach synthesis, because without synthesis there is no science. Ultimately, we want to understand the thing as a whole. Rational analysis is necessarily violent, but we need it if we are to move beyond purely descriptive operations incompatible with true knowledge of dynamic facts and living things. The seductiveness of interdisciplinary efforts emerges precisely from this challenge. They unite diverse and apparently disparate knowledges. This is an endless test because knowledge is constantly growing and diversifying.

Constructing a coherent whole endowed with an internal logic through the accurate manipulation of elements collected in this way is even more difficult. Yet, without this internal logic, any interdisciplinary effort to advance geographical theory will not exceed a mere catalogue of citations or a list of comparisons underpinned by analogies. Traditional general geography cannot become a true theory. There is a fundamental opposition between that old school of thought and a fresh and renewed general geography.

The search for more advanced theoretical process is possible only through abstraction, which can be achieved only through the intermediation of the categories that define a given reality. In our time, when each concrete fact is the result of many determinations, the mere apprehension of the fact in itself is insufficient for deduction. The fact is only an example; one thing among others. We need a systematic process that emerges out of ideas that are to some extent independent of the examples which underpin them.

We can turn again to Whitehead (1938, 196): "the topic of each science is an abstraction taken from the concrete and complete functioning of nature." Geographic space is nature modified by man through labor. The idea of a natural nature, in which man either does not exist or is not central, cedes its place to the idea of the permanent construction of artificial, social nature, synonymous with human space.

An Effort to Define Space

I do not mean to be unfair—I understand why geographers have spent more time defining geography than space. The latter is extremely hard to do. The same could be said of space as Saint Augustine said of time:

"if I ask myself if I know what it is, I respond that I do; but, if I ask myself
to define it, I respond that I do not know how."

As an object of concern for philosophers from Plato to Aristotle, the
idea of space still covers a broad range of objects and meanings. The
everyday utensils of domestic life—an ashtray, a teapot, a statue, or a
sculpture—whatever their dimensions, are space. A house is space; so is
a city. There is the space of a nation—the synonym of the territory, of the
state. There is terrestrial space, the old definition of geography, the crust
of our planet. And there is extraterrestrial space, recently conquered by
humans, and the continuing mysteries of outer space.

The space that concerns us is human or social space, that which contains or is contained by all these different types of space. Indeed, these are
the objects of particular disciplines that define them in particular ways:
semiotics, sculpture, painting, urbanism, physics, astronomy, and so on.
But the dimensions of these other spaces matter little if their content is
directly legible to human sensibility. Defining a vase of flowers, a sky-
scraper, a planet, or a constellation is easy. The mind is quickly satisfied
with such definitions. But when our curiosity turns to human space, things
become more complicated because it is our dwelling place. It is where we
live and work. Its forms and content are incredibly various. Including
such a great multiplicity of facts within a singular definition is immensely
challenging. Everyday language and conceptual language are full of meanings that correspond to particular types of space. What, therefore, is the
space of the human? It is geographic space, you could respond. But what
is geographic space? Defining it is hard because once geographic space is
social space, it has a tendency to change with the historical process.

To establish a valid debate, the first question we must ask is whether
we can find a single definition for this category *space*. Or do we have two
different things to define: first, space as a permanent category, or *space—
eternal space*—and, second, space as it presents itself to us today—*our
space*, the space of our time.

Space as a permanent category would be a universal form with permanent relations between logical elements. These would be identified through
researching the immanent: that which passes through time and not that
which belongs to a given time and place, the properly historical, transitory
fruit of a delimited topographical combination, specific to each place. The

concept of a social system traverses the idea of this time and place. It is fundamental to *our space,* the second type of space to be defined. In both cases the definitions cannot be immutably fixed and eternal.[6]

For space as a universal and permanent category, scientific and philosophical progress means we can define it differently in each moment. The natural sciences are not exact. At each historical moment, so-called natural phenomena are defined differently depending on progress in "exact" sciences and in the sciences of knowledge, philosophy, and society. In terms of space as a historical category, "it is the meaning itself of the objects, of their content and the relations between them which change across history."[7] As Ludwig Feuerbach argued, the social world around us is not eternal. In fact, the two paths intersect. Knowledge of space as a universal category is included in knowledge of space as a historical category, and vice versa. The interaction between universal laws and individualized historical behaviors helps construct a concept (if not quite a definition) of space that is no less philosophical for being operational.

The idea of place—a discrete portion of total space—has preceded the concept of space: Aristotle formulated this idea and Einstein insisted on it (Jammer [1954] 1969, xiii): "it seems [the concept of space] was preceded by the psychologically simpler concept of place." Place is, before anything else, a portion of the face of the earth identified by a name. What makes "place" specific is a material object or a body. A modest analysis shows that a "place" is also a group of "material objects." But if, from a purely psychological point of view, the concept of place occurs to us before the concept of space, from a theoretical and epistemological point of view, the concept of space precedes the concept of place.

One of geography's philosophers, William Bunge (1966, 125–57), emphasizes that the universe is not a mass of things but a system made up of systems that interact between themselves as if they were simply elements. What happens in one place depends on the totality of places that make up space. Was this not the epistemological principle of the Arab historian-geographer Ibn Khaldun? Was this not also, later, the basis of the thought of Leplay and the foundation of the general principles of geography of Vidal de La Blache? It is to this last geographer that we owe the idea of the unity of the earth. In their geographical studies Demangeon, in France, and Chauncey Harris, in the United States, took this as the basis of international reality. Referring to urban studies, T. G. McGee (in Jakobson and Prakash 1971, 160) recently argued that "cities are only parts of a total social and economic system which is not only national but international."

Space has to be considered as a set of relations. These relations are brought into being through functions and forms that present themselves as evidence of past and present processes. Space is a set of representative forms of past and present social relations and a structure represented by social relations which manifest through processes and functions. Space is, therefore, a field of unevenly accelerating forces. Hence spatial evolution does not happen in the same way in all places.

Once again, the idea of relativity introduced by Einstein seems fundamental. It substitutes the concept of the material for the concept of the field. This presupposes the existence of relations between matter and energy. It is a crude comparison, but we could see forms as matter and energy as social dynamics.

11

Space
Reflection of Society or Social Fact?

There are some philosophers for whom what exists is nothing more than the creation of our minds. Ferdinand Gonseth's ideas about space, like George Berkeley's, are often put into this bracket. Benedetto Croce ([1915] 1968, 73) is even more explicit. For him, "a fact is historical in so far as it is thought, and . . . nothing exists outside thought."

A Form of Perception?

According to Karel Kosík's ([1963] 1967, 60) interpretation of F. Gonseth, in the act of knowing man puts himself not in relation to nature itself but in relation to particular horizons and images that are historically mutable and can capture the fundamental structure of reality. For Gonseth (1948, 413), "the natural world is constituted in such a way, and we ourselves are constituted in such a way, that reality does not allow us to achieve a definitive knowledge of its true essence." For Bergson (see Russell 1945, 798), similarly, space cannot be "real"; it cannot itself be a bearer of existence. Therefore, there are in fact no things: things and states are visions that our mind apprehends.

According to Jim Blaut (1971, 18), this interpretation would have been espoused by both Ratzel and Hettner. For the latter, space would be considered as *Anschauung*, a way of seeing things or an intuition. In his "Das Wesen und die Methoden der Geographie" (1905), Hettner argued that "space is only a form of perception."

However, there is a big difference between declaring that space is a form of perception and affirming, like Russell ([1948] 1966, 234), that "the unitary space of common sense is a construction, though not a deliberate one." He was referring specifically to the representation of space within each person's mind.

However, according to Ernst Cassirer (1953, vol. 3, 145), we comprehend the space of things—physical space—on the basis of the evidence of our five senses, the comparison of this evidence, and the construction we can make on its basis. The space thus produced is a schematic intellectual production, or pure geometrical space characterized by qualities such as "constancy, infinity and uniformity."

As for human space, that's another question.

Hegel and Space

For some, such as Shlomo Avineri, Hegel, the founder of the modern dialectic, is among those for whom space exists, above all, in our thoughts. He argues that Hegel's inclusion of inanimate nature in his dialectical system is irrelevant because for him, nature constitutes a form of self-alienation (1968, 65). In the Hegelian system, "nature thus did not appear as a subject but as a mere predicate of thought, and the self-reflecting spirit must emerge from abstraction and become objectified," writes Avineri (1968, 11–12). The objectification of this spirit is achieved precisely through the intermediation of nature.

Avineri's reading of Hegel is also, in certain respects, a rereading of Marx and his interpretation of Hegel. Hegel admits that the creation of man is accomplished through the modification of his relations with nature. Marx refutes this idea. He argues that Hegel spiritualizes man and nature and reduces history and life itself to the level of a concept. August Cornu (1945) reproduces Marx's idea that "to arrive at an exact concept of man, of nature, and of their relations, we must consider them in their concrete nature."[1]

In the chapter dedicated to the geographical foundations of history in his *Philosophy of Right*, Hegel refers to the soil, the climate, and the geographical situation. He writes that "the understanding of right passes through the analysis of its contents in which, besides the particular national character of each people and their own stage of historical development, includes the total complex of relations that have the necessities of nature as their basis."

In *Reason in History* (1955, 187), Hegel wrote that "the natural context is simply the geographical basis of universal history and not, in the first place, the objective precondition of social labour, though the relations of labour can be treated as a reflection of the natural context."[2]

Hegel would also have admitted that nature exists for itself, as an object: "The sun, the moon, the mountains, the rivers and natural objects of all types which surround us *exist*" (1942, 1962, 166).[3] "When we use a tool, or derive water power from a stream, we recognize and do not annul the particular character of the object that serves our purpose; ie we recognize it as an *object* and to that extent as self-subsistent" (Hegel 1962, 348, n.146). Yet, "nature, as the creation of God," ceases to exist in an autonomous form and depends on the Idea for its philosophical construction (Hegel, 1942, 1962, 348).

Following Feuerbach, Hegel sees nature as the objectification of the spirit. But Avineri (1970, 12) interprets this as if Hegel's position is one of limitless abstraction.[4]

Space, a Reflection?

In his *Phisiologia* (Paris, 1637), Campanella argues that God created space as a "capacity," a receptacle for bodies. "*Locum dico substaitam primam incorpoream, inimobilem, optam et receptandum anne corpus.*" This is close to Kant, when that German philosopher-cum-geographer, in his *Critique of Pure Reason,* considered space to be a "condition of possibilities of phenomena."

For W. E. Moore, more recently (1963, 8), space is a condition of behavior, but a passive condition that shifts when human behavior shifts. For Moore, only time is intrinsically dynamic. Space gets its dynamism only from changes in social values, interests, and technologies.

Many modern and classical authors argue that space is a reflection of society, a backdrop on which social conditions are readily inscribed, as they happen. But not everyone agrees.

"The city is a projection of society on the ground," writes Henri Lefebvre in his polemic book, *The Right to the City* (1996, 109). This sentence, taken in isolation, has allowed others to fill in the gaps and to distort it into the idea that "space is a mirror of society." But this is a very different position. We will begin by discussing this second version.

The issue is to find legitimate and adequate terms to answer Paul Vieille's question: "Is spatial organization only a reflection, or a projection of a social organization that is defined independently of it, and in an autonomous manner, or does space intervene (and how?) in the historical process?"

Responses fall basically into two camps: (a) Kantian space, "an *a priori* representation, necessary basis for external phenomenon"; (b) space as a reflection of society. In the first group, we have the notion of *space-container*.[5] In the second group we have the idea of a space that merely mirrors social phenomenology. In both these hypotheses, space is considered not as a structure or order endowed with relative autonomy but as a *level* of society, by virtue of being a *reflection* of other structures, subsystems, or orders, whose facts space synthesizes. This is a misinterpretation, because whatever the subsystem or the social structure, space equally synthesizes all the givens of society as a totality.

When we consider space as a mere *reflection,* we put it on the same plane as ideology, even if we do not mean to classify it as a structure.[6]

This idea of space-as-level is another product of the philosophical inheritance of Kant and Newton but also of positivism, which even Marxists could not escape.

The truth, however, is that space is far from being a "neutral framework, empty, immense, in which the living being can produce itself." This is the image going back to the sixteenth century that Charles Morazé justly criticizes (1974, 118). As a philosopher of history, Morazé argues that space was for a long time taken to be a "mathematical void," but, in the age of vitalism, it would come to be considered to be a reflection of time. But what kind of vitalism has taken up this hot philosophical topic of the nature of space? Claude Bernard's? Perhaps. But the idea that Leibniz put forward—of space as a system of relations—and the other that François Perroux developed—of space as a field of forces—are the precursor and the result, respectively, of Einstein's concept of relativity. This concept re-poses the problem in new terms because, whether the individual perceives it or not, the "system of relations" and the "field of forces" act beyond the individuals who are subjected to them and independent of their individual choice.

A Social Fact?

Should we, therefore, apply Durkheim's important notion that ([1895] 1962, 14): "the first and most fundamental rule is: *Consider facts as things*"? Can we assimilate geographic space to the definition of the social fact as "every way of acting, fixed or not, capable of exercising on the individual an external constraint", or even "every way of acting which is general throughout a given society, while at the same time existing in its own right independent of its individual manifestations"? Durkheim proposes, on the other hand, that the expression "ways of acting" is connected to another, the "ways of existing." For him, the means of existence are the crystallization of the means of action. He gives us an example (1895, 1962, 12) when he alludes to "the type of habitation imposed upon us."

According to this Durkheimian meaning, space is, therefore, a thing; it exists beyond the individual and imposes itself as much on the individual as on society as a whole. Therefore, space is a social fact, an objective reality. It imposes itself on individuals as an outcome of history. They can each have different perceptions of it, proper to the relations between subject and object, but the individual perception of space is one thing and its objectivity is another. Space is neither the sum nor the synthesis of individual perceptions. Being a product—that is, a result of production—space is a social object like any other. If, like any other social object, it can be understood through multiple kinds of fixings and gatherings, this does not empty it of objective reality.[7]

In the final analysis, the *reality* of a city, a cultivated field, or a road is *the same for all individuals*. The reality of each individual enables him to see things from a particular angle. But, as the result of human labor—an artifact—space retains its objective character throughout its own transformations. The basis of knowledge and the interpretation of spatial reality cannot, therefore, be found in the sensations or in perception. This basis is false and without substance. We can understand space only through the production of space.

Comte proposed that social phenomena be considered as natural facts and subordinated to natural laws. He made a double mistake. First, he assimilated the laws of the functioning of society to the laws of the physical world.[8] Then he took the ideas themselves, and not the things, as the material of study.[9] This was the critique Durkheim justly leveled against him ([1895] 1962, 19). He thought that the positivist concept of social evolution was subjective because the linear evolution proposed by Comte ignored the specific, concrete, and objective evolution of individual societies.

Social relations themselves can be studied as objective. According to Norman Geras (1971, 641), "the fact that the material forms of capitalist relations are not natural does not deprive them of their objectivity, that is, of their character as independent objects in relation to the social agents who rule according to their own laws, and whose origin and explanation cannot be attributed to human subjectivity."[10] Silvano Sportelli's (1974, 91) observation is fitting; he says that *social objectivity* is frequently reduced to a *natural objectivity*. This is the same as forgetting that nature is the object of permanent transformation thanks to human activity and that nature is therefore a social reality and not exclusively natural. In this sense, the word "natural" must be taken as a synonym for "social," in the same way that the words "nature" and "space" can be assimilated into each other. To acknowledge that space is a social fact is to deny that it can be interpreted outside the social relations that define it. Many phenomena presented as if they are natural are, in fact, social.

Socialized nature, then, names what geographers normally call "space" or "geographical space."

Space is a social fact in the sense in which Kosík (1967, 61) defines social phenomena: a historical fact insofar as we recognize it as an element of a whole. These perform dual functions: defined by the whole but also defining it; simultaneously producer and product, determinant and determining, a revelatory force deciphered by those who reveal it, as it acquires its own authentic meaning, it renders meaning onto other things. According to this interpretation, space is a social fact, a social factor, and a social order.

12

Space
A Factor?

"Localisation economies tend to heighten the influence of urbanisation economies" (Bergsman et al. 1972, 264). "The principle of accumulation teaches that when the action of the market is free, a group of people, *a city or a region of a country* that thanks to its particular circumstances finds itself in a historically dominant position tends to have its position reinforced. The groups, people, regions or countries that fall under their dominion remain static or, in the most marked cases, excluded from the process of accumulation" (Marrama 1961, 79).

The Reproduction of the Spatial Template

The organization of space tends to reproduce its principle lines of force. If we examine, for instance, maps of the distribution of human settlement during the four and a half centuries of the modern history of Venezuela, we see that the marks of human presence in the territory repeat themselves, albeit with some nuances. As is natural, the character and experience of settlement changes both qualitatively and quantitatively, but the roots of settlement exert a powerful influence over what follows.

Similarly, the original plans of cities like Paris or London reproduce themselves at greater or lesser scales across time. Changes from different periods do not completely erase the city's original morphology.

Writing about Asia, Gerald Desmond (1971) shows that it was in the period of European domination that the urban network that now dominates the region was established. African countries, but also those of other underdeveloped continents, find it difficult to change the model of public spending and alter the geographical distribution of investments. Their influence is therefore compounded. Referring to Zambia, L. S. Chivino (1973) demonstrates the relationship between actual cash spending and spending envisaged in the national budget. It shows that actual cash spent is greater in the more developed zones that already have basic infrastructure. This is to the detriment of other regions, particularly the least fortunate ones. Good intentions in plans and budgets cannot resist the force of facts, ordered by an economic and social structure that reproduces and reaffirms itself. In Brazil, in spite of efforts to retain population in the interior of the country, the tendency to reproduce models of distribution is even greater.[1]

The construction of modern means of circulation is an example of this spatial inertia. Roads are constructed parallel to railways, motorways follow the routes of old roads, and bridges are replaced in place, even if natural conditions are not the most amenable. Any number of other examples could show the force of historical localizations in the present.

In East Africa, British colonization deliberately favored particular sections of territory. They were intended as galvanizing poles for the rest of space. In these zones, even after the independence of Kenya, Uganda, and Tanzania, the legacy of infrastructure and inherited activity attracts new investments, giving these parts of the territory an enviable advantage over the rest. This situation particularly affects Tanzania. The government, currently attempting a modernizing socialist project, wants to reduce the influence of the Arusha-Moshi region, which is still today economically incorporated into the Nairobi-Mombasa axis, linking the two principle cities of Kenya. This axis was destined, according to the British plan, to play a commanding role in the (now fragile) East African community.

As transport infrastructure is best in the privileged zone, it is used more than elsewhere. The rate of profit of transport businesses is therefore higher, which is positive for business as a whole. As a result, a great number of other activities arise that rely on more intense circulation. Businesses develop much faster, and agricultural production is stimulated by the greater circulation of products. The fact that people circulate more easily stimulates business and ensures a clientele for transport industries. Thus the most dense and important part of the country reinforces itself, obliging the central government to invest to restrengthen the trend.[2]

The Mobility of Capital Is Relative

Capital itself does not possess the mobility often attributed to it. This is even more obvious in underdeveloped countries, where only certain places can offer the conditions of profitability that capital requires. Writing about developed countries, R. C. Estall (1972, 196) notes that "even for the large enterprise, the freedom to dispose of new investment funds at places where returns would be highest can be heavily constrained by the need to support existing capital investments."[3] In underdeveloped countries, big businesses, in order to be profitable, locate themselves in metropolitan regions. As well as economic and social infrastructures, these regions offer better economies of scale and greater ease of interpersonal and long-distance communication. The presence of cheap labor is another draw. They do create enclaves, but above all for the production of raw materials for export to richer countries.

Space in the Social Totality

Doubtlessly, the ever messier search to maximize profit in this phase of the expansion of the capitalist system benefits some locations and denigrates others. Sorre (1957, 66–67) gives the example of the significance of already worked land, and how "the permanence of this land dominates the rural group," constituting an influence that is no less important by virtue of being relatively straightforward. "The land is not a witness, or a mute actor." At the other extreme, large cities are also an example of permanence, based on ever more powerful economic, political, social, and cultural laws, as proved by the cumulative and irreversible macrocephalic growth since the beginning of this century.[4]

The way this process of top-heavy growth currently unfolds in underdeveloped countries results from technological progress and its concentrating tendency. Initially favored cities benefit from a selective accumulation of advantages and accept new enhancements.

The presence of a constantly growing population ensures that a good part of the overhead capital[5] and infrastructure necessary for desirable economic activities to establish themselves is available. Beyond this, the concentration of public investments at certain points in space generates the tendency for the coefficient of capital necessary for the installation of new activity to rise (Dasgupta 1964, 180–81).

Once established, this position of dominance continues to deepen even if other centers also undergo significant growth. "As soon as industry and trade become concentrated in a particular centre, they themselves give to that centre an advantage for further development" (Hicks 1959, 163). Thus, we can speak of an immobility of advantages resulting from agglomeration. This is an enduring immobility. Thanks to cumulative developments (Rémy 1966, 69), advantages are realized in the place where the installation first occurred.

Instead of "top-heavy growth" it would be better to say "metropolitan region." The cases of São Paulo, Mexico City, Buenos Aires, Caracas, and many others are clear. But the phenomenon also reaches other dynamic regions that are not only metropolitan, for instance in certain mining zones and even agricultural regions from which some countries of the Third World derive the essence of their strict divisions.

From the moment the movement begins, it becomes irreversible. The development of economic activities that are considered fundamental (and morally and politically legitimate) necessitates the accumulation of an unprecedented volume of economic and social investments greater than those of the rest of the country. It is natural that these infrastructures attract others, whether thanks to the prediction that existing activities will grow or whether because other activities are already in place. The country is obliged to dedicate an increasingly substantial part of its budget and resources to rich zones.

The Central Region and the Copper Belt of Zambia are good examples. These provinces, on which the budget envisaged spending a significant proportion of its investments, in practice far exceeded even these planned expenditures. While in the Central Region and the Copper Belt real expenditure exceeded 157 percent and 138 percent of planned spending, respectively, in other zones the total invested settled between 60 percent and 73 percent of envisaged spending. Urbanized provinces got 52 percent more than envisaged, while rural areas got 21 percent less (Chivino 1973, 197). The advantages denied to other parts of the country constitute a permanent invitation to investors. "Just as the economies of agglomeration are made the most of by the capitalist sectors of the respective cities, so the diseconomies are assumed by the State and by the population" (Funes 1972).

On the other hand, national economic metropoles benefit from a strategic position in the modern transport network. Experience shows entrepreneurs that investing outside the centers of growth is, at best, minimally profitable (Johnson 1970, 150).

We can take Jakarta as an example. According to S. V. Sethuraman (1974, 3), the annual rate of investments is 32 percent of the national total, but this has been growing in recent years. Again, this is the result of the tendency to concentrate investments, above all in the Third World. "Localisation economies tend to heighten the influence of urbanisation economies" (Bergsman et al. 1971, 264):

"The principle of accumulation teaches that when the action of the market is free, a group of people, *a city or a region of a country* that thanks to its particular circumstances finds itself in a historically dominant position, tends to have its position reinforced. The groups, people, regions or countries that fall under their dominion remain static or, in the most marked cases, excluded from the process of accumulation" (Marrama 1961, 79).

Perhaps it is superfluous to insist that this is a general rule and that efforts at demographic dispersal and industrial decentralization that have been attempted up to now never had a future. Of the many projects proposed to revalorize medium-size cities, the most successful delivered only ephemeral statistical effects. When comparing rates of growth of medium-size cities to the economic metropole, people forget to put such comparisons in the context of the social formation. The numbers do not show the most important flux—surplus value. This is the single most important factor directing what large cities retain or—more frequently—what they send overseas.

The Role of Roughness

At the end of the last century, Friedrich Engels (1963, 410) already considered geographical space to be an element in the formation of society. In a letter sent to Starkenburg (25 January 1894), he explicitly included "under economic conditions . . . the geographical basis on which they operate and those remnants of earlier stages of economic development which have actually been transmitted and have survived."[6]

This principle is also universal in time. In relation to *kinship*,[7] Sandra Wallman (1975, 340) shows that "the logic [governing relations between kinsmen] varies with context within a society as readily as with changes in socio-geographic settings."

Still, the role of space often passes unnoticed or unanalyzed in all its depth.[8] Like Sartre on materiality, we should ask ourselves why no one has "attempted . . . a study of the type of passive action which materiality as such exerts on man and his History in returning a stolen praxis to man in the form of a counterfinality" (2004, 124).

Space is worked matter *par excellence*. No other social object has such dominion over men or is present in such a form in the everyday life of people. The house, the place of work, meeting points, and the routes that unite these locations are passive elements that condition the activity of men and order their social practice. Praxis, the fundamental ingredient in the transformation of human nature, is a socioeconomic fact, but it is also a tributary of spatial conditions. As Roger Caillois said (1964, 58), space imposes a set of relations on each thing because each thing occupies a certain place in space.

We can cite Sartre again: "the practico-inert" robs human action and imposes a "counter-finality." When dealing with human space, we can speak no longer of the practico-inert but rather of dynamic inertia. Representation is also action, and tangible forms participate in the process as much as actors themselves (Morgenstern 1960, 65–66).

Claval (1970, 120) pinpoints the problem when he says that the formula according to which "human geography is the study of the projection of societies on the face of the earth" runs the risk of being misunderstood because the relationship between sociology and geography is not univocal. The spatial equilibrium of a society is the projection of its multiple dimensions in concrete space, but the restrictions that this imposes on it have repercussions for its whole structure. Human geography is not a simple application of the social sciences; it is not another one of them but constitutes a facet of their multiple aspects.

Manuel Castells (1977, 127) speaks of the "persistence of ecological spatial forms created by earlier social structures." Where Castells speaks of "ecological forms" I prefer to use a geomorphological term: *roughness*. Ecology works with durable or ephemeral, natural or social forms, that is, those introduced by man. Roughness is constructed space, the historical time that transforms a landscape, incorporated into space. Roughness offers us, even without immediate translation, the remains of an international division of labor, manifested locally by particular combinations of capital, technics, and labor.

Thus, space—landscape-space—is the witness of a moment of a mode of production in its concrete coming into being, the witness of a moment of the world.

The mode of production that creates fixed spatial forms—through the intermediation of its determinations (different determinations in the same place at the same time)—can disappear without the fixed forms themselves disappearing. The moment crystallizes in memory, as Lefebvre would say (1958, 345).[9] And, to repeat Irvin Morgenstern, as the memory of a present that is past.

Space, though, is a witness. Through the memory of a constructed space, and through fixed things in the created landscape, it witnesses a *moment* of a mode of production. Thus, space is a lasting form. It is not unmade as processes change. On the contrary, some processes adapt themselves to preexisting forms, while others create new forms to insert themselves within.

Modes of production manifest through *means of production*. Their longevity is only known a posteriori, but they can overtake the mode of production at one or various *moments* or even for its whole existence. This is the case of European buildings of the Middle Ages: castles, cathedrals, streets. . . . Modes of production follow one another, but the social objects they created remain solid and even retain a function in production.

Thus, when a new moment—a moment of the mode of production— replaces another, it finds preexisting forms it must adapt to if it is to realize its own (spatial) determination. So space is an effective and active condition of the concrete realization of the modes of production and their moments.[10] Geographical objects appear in places that correspond to the objectives of production in a given moment. Their presence influences subsequent moments of production.[11]

Man works through what he inherits.[12] As Feuerbach wrote, "the sum of productive forces, capital funds and social forms of intercourse . . . every individual and generation finds in existence as something given."[13]

Castells (1976) argued that "society cannot be 'reflected in space' since it is not external to space" (70) but that to "analyse space as an expression of the social structure amounts, therefore, to studying its shaping by elements of the economic system, the political system, and the ideological system" (1977, 126). While space means many different things, there is an ambiguity we need to resolve—which Castells hasn't quite clarified—between landscape and actual space. They are both real but not synonymous.

This ambiguity does not come from forgetting temporality. Walter Isard, for example, prefers to say that space is the result of the superimposition of social systems (Isard 1960, 85; Isard et al. 1968, 75). Mistakes arise, rather, from not distinguishing landscape from space as such. This distinction is key if our theoretical position is to be based on analysis of actually existing facts.

Without an analytical approach that distinguishes between the constitutive elements of total space and without considering temporality, we cannot apprehend either space as a perpetually evolving real object or its relations with society in permanent motion.

If we adopt an analytical and dynamic perspective we can see, like Roberto Briceño et al. (1974, 34), that space is not "innocent": it is in the service of social reproduction.

13

Space as Social Order

Definitions of society ordered as a system or a structure (or even as a totality) often exclude space. Odd though it may seem, in this respect Marxist and "bourgeois" theories are similar.

Talcott Parsons and Neil Smelser (1956), for example, divide the social system into four subsystems: economic, political, integrative, and the maintenance of standards. Space is not considered. In Marxist thought, there are nuances of classification, but there is a similar lack of reference to space. In his book about fundamental questions of Marxism, Plekhanov (1907, 117–83), who exaggerated the role of nature in the orientation of social life, distinguished five levels as indispensable to the definition of society:

1. The state of the *productive forces*; 2. The *economic relations* that these forces condition; 3. The *social and political regime* that rests on the economic "base"; 4. The *psychology of social man,* in part determined by the economy, in part by the whole of the social and political regime that it builds; 5. The *diverse ideologies* that this psychology reflects.

According to Jakubowski (1971, 96), within the term "social psychology" Plekhanov includes "the conscious, general reaction to the social relations of an epoch" that are manifested "in distinct, concrete ideologies." Jakubowski therefore proposes simplifying the Plekhanovist schema and distinguishing only three "regions" (Louis Althusser and his followers often employ similar terms). The elements of society, therefore, would be limited: (1) the economic base, itself determined by the productive forces; (2) the corresponding political and legal order; (3) the ideological superstructures that rise from this edifice.

That construction is only slightly different from what is offered by other Marxists. Charles Bettelheim (1970, 1445), for example, writes that "the set of social relations of production and the ideological and political relations constitute a complex structure whose elements are reciprocally one another's 'cause' and 'effect,'" or, more strictly, they "mutually sustain one another." There are no references to space. Almost identically, P. L. Crosta (1973) writes that "society is formed of the complex of political, legal, economic and productive structures." Gianfranco La Grassa skips space too: "the economic formation of society"—in the sense of "structural make-up"—"is constituted by subordinated economic and social forms, beyond the dominant mode of production" (1972, 107).

For Harnecker (1973, 147), the social formation is a complex structure composed of complex regional structures (economic, ideological, legal, political), all articulated on the structure of the relations of production. She promotes studying "each regional structure in its relative autonomy and in line with its own characteristics." But the list of so-called regional structures includes, exclusively, the economic, the ideological, and the legal-political. Once more there is no mention of space as part of social order.[1]

Córdova (1974), for whom modes of production are a particular form of the modification of nature, makes the same omission. He outlines a process for studying the specificity of relations internal to a given mode of production. His "subsets" include the three classic orders. In a distinctive vocabulary, he writes of "technical relations of production (techno-economic structure), social relations of production (socio-economic structure), political and legal relations (legal-political structure), ideological and cultural relations etc." Perhaps space is within this "etc."

This is certainly not the only unusual proposition. Cohen (1973, 13–14), the Argentine sociologist, argues that the social structure is made up of three systems: production, stratification, and domination. Each of these systems is continually changing at different rhythms and intensities. Their asynchronicity comes from the relatively autonomous functioning of each system and constitutes "a fundamental fact for the understanding of the social structure." This adds a new element to understanding social evolution: the uneven and combined development of structures in motion. Precisely for this reason, space must be but is not in the principal scope of any analytical schema. "Spatial structure" does not evolve at the same rhythm, or in the same direction, as other orders of society.

The classical schema is so entrenched that even work by radical researchers on the relations between the social formation and space cannot escape its gravitational pull. In one such plan (Michelena, July 1973), happily later improved, the traditional classification was retained. Space was considered "social spatial" or "historical social" and put outside the main set of social orderings. As I will now try to show, this approach, like the previous ones, is insufficient.

A Social Structure Like the Others?

"Should we regard urbanism as a structure which can be derived from the economic basis of society (or from superstructural elements) by way of a transformation? Or should we regard urbanism as a separate structure in interaction with other structures?" This is Harvey's question (1973, 293), but he immediately proceeds to "leave these questions aside for the moment, for they will be the foundation for the second part of this conclusion."

I agree with his fundamental theoretical position, but the way Harvey puts the problem is still similar to Castells's formulation—both describe the urban system as a "social structure."

Yet the problem is really much broader: it is not urban space that constitutes a social structure but human space as a whole. This pushes us toward a fundamental theoretical and methodological exercise: demonstrating the real place of human space in global society or, better, in the economic and social formation.

First, a note of caution: we must not confuse the functional and systemic qualities of phenomena with the objects that correspond to them. Thanks to its structure, more than to its form, space's functional qualities— like those of any other social structure (or level of society)—are a reflection of global society. Its dynamism comes from the cleavage between global society itself and its distribution across territory. In this sense, space is a social fact that imposes itself on everybody. But seeing the systemic qualities of space gives it new attributes, including the capacity to condition and even to some extent determine (conditionally) the evolution of social structure.

Is this enough for us to see space as a structure of society on an equal footing with other social structures? Could others respond that space is only a social fact, a concrete phenomenon that makes itself felt on all members of a society, not on society itself?

Our primary interest is in the characteristics of social structure and in ascertaining whether its attributes can also be found in space. If so, we can include space in the list of social structures.

Space, like other social orderings, tends to reproduce its principal outlines. Spatial structure—space organized by man—is, like other social structures, a subordinated-subordinating structure. Like other orderings, though contingent on the law of the totality, space has an autonomy that manifests through its own laws, specific to its own evolution.

These are themes that are now being discussed systematically. V. V. Pokhishevskiy (1975) has recently outlined the influence of spatial forms on social processes and critiqued the contrary argument.[2]

In fact, space cannot be only a reflection of the mode of production because it is the memory of past modes of production. In its own forms, it survives the passing of modes of production and their moments. This characteristic of Sartre's *practico-inert* turns against its creator and is the basis of the existence of space as a social structure, capable of acting on society itself and reacting against other structures of society. Social determinations cannot ignore the concrete spatial conditions that pre-exist them. A new mode of production, or a new moment of a continuing mode of production, cannot make a clean slate of preexisting spatial conditions.

A Subordinated Structure?

Space does not depend only on economic structure.

If the spatial is subordinated to the economic, the first question is whether the economy can function without a geographical base. The answer, obviously, is no. Even if the word "geographical" is taken as a synonym for natural conditions, in the way that many economists and other social scientists wrongly understand space, the answer is still no.

In an otherwise laudable essay, the Marxist François Ricci (1974, 131) argues that "the scientific elaboration of the economy does not dislocate or reorganize the natural facts on which economic activity is built."

Falling into the trap of a narrow definition of "geographical," Ricci pushes a dualist conception of the relations between productive man (economic activity) and nature (natural facts). Explaining Marx through the logical structure of *Capital*, he ends up encouraging readers to reject Marxism.

The economic is presented as a complex social reality because it is a field of activity oriented toward the production, distribution, and consumption of material objects. It is oriented by the mechanisms of production, distribution, and consumption but is just one particular aspect of all noneconomic activities. On its own, it "does not possess the totality of its meaning, nor its finality, but only a part of it" (Godelier 1969, 31).[3]

Neither can we deduce the structures of society from the economic infrastructure: "the economic structure does not automatically produce anything" (Harnecker 1973, 147).

Marx has played an involuntary role in economistic interpretations of social relations. Engels wrote in a letter to J. Bloch (21–22 September 1890):

> To some extent you could blame Marx and I for the fact that the youth sometimes give more weight to the economic aspect than it merits. In the face of our adversaries we tend to underline the essential principle that they deny, but then do not find the time, place nor occasion to show the true value of the other factors that participate through their *reciprocal action*.

The backdrop to Marx's thought is, nevertheless, the totality. This is clear in his famous *Introduction of 1857*. It is possible, of course, as Althusser does (1965, 9), to cite Marx to conclude that there is a dominant structure (*structure à dominante*) responsible for the articulation between the parts and the order of the whole.[4] György Lukács (1971) reminds us that the thesis of dominance does not prove, but contradicts, the idea of the totality. For Lukács, the category of the totality, an inheritance left to Marx by Hegel, consecrates "the all-pervasive supremacy of the whole over the parts." It constitutes the essence of Marx's method and the "foundations of a wholly new science." Primacy corresponds to the totality as structure, which is above its substructures and exceeds the temporal succession of the various categories.

Córdova's (1974, 154) position is crucially different from Althusser's. Though conceding, like Althusser, the dominant character of the social structure "in the last instance," Córdova emphasizes the socioeconomic structure, not the economic structure *tout court*. For him, this socio-economic structure would introduce "a specific order in the articulation of different structural planes and their mutual relations." Córdova clari-fies that "each of these planes has a *relative autonomy* in their historical movement and, in the same way, has a relative capacity to influence other planes, including the *dominant* structure."

When Castells writes that space is a subordinated structure, we need to remember that no dialectical relation can exclude the action of one of its components. We refuse to countenance the idea, therefore, that there are subordinate structures whose transformation is due only to economic determinations.

Organized space is not a social structure that depends only on the economy. If this may have been the case in the past, today it is clear that other influences are involved in changes in the spatial structure. Political facts, for example, are a driving force, for instance when the state, for doctrinal or conjunctural reasons, decides to reorganize its territory to better secure its sovereignty. For motives that the rest of civil society is excluded from, the security forces of a state can suggest or demand that a government populate frontier regions or build strategic roads, ports, or airports. A concrete example is the populating of Amazonia by countries in its basin. This is surely a typical case of international policy being driven by contemporary political exigencies. In such instances, instruments of production are created, even if they have no correlation with the need to produce. But such resources might be called upon, now or in the future, to exercise certain functions in the productive process. However, even before such "dormant capital" (Santos 1975b) comes to play a role in the produc-tive process, a spatial process has begun. Changing the total distribution of the instruments of production modifies the relations between produc-tive forces and relations of production in the space as a whole.

The Specificity of Space

Organized space is also a form, an objective result of the interaction of multiple variables through history. It has *dynamic inertia*: forms are as much a result as a condition of processes. Spatial structure is active, not passive. Like other social structures, its autonomy is relative.

This active, or dynamic, inertia has polyvalent manifestations: the attraction of large cities on the potential labor force, the magnetic force of capital, the superabundance of services and infrastructures whose uneven distribution reinforces inherited dimensions.

Using the case of Venice, Giorgio Ferrari (1974, 85) saw space as underpinning the model of development, producing rents, and conditioning the continued differentiation of the labor market. Thus, space has a fundamental role in social structures and collaborates in the reproduction of social relations. As Donatella Calabi and Francesco Indovina (1973, 18) argue, "the organization of territory is not only a variable of, but, to a certain extent, a fact of the capitalist process itself."

Surveying the spatial conjuncture, whether in the city or in any other fraction of total space, we reach similar conclusions. The active role of space in social evolution is undeniable. As Vieille (1974, 30) wrote: "when we consider economic and social processes, space is, in fact, a dimension of the mechanisms of transformation, of the practice of social groups, of their relations; it contributes to producing, reproducing, and transforming the modes of production. Space is, therefore, an active dimension in the becoming of societies."

For all this, is it true that space has its own, specific, exclusive role in processes of change?

If, with each transformation of the whole of social relations, space went along with changes in other social structures and adapted itself immediately to the necessities of *optimum* functioning, it would have nothing but a passive role. But the dynamic inertia that space is endowed with ensures its tendency to reproduce the overall structure from which it originates. Yet it necessarily mediates social reproduction, altering initial objectives or driving them in a particular direction.[5]

The specific role of space as structure of society comes, among other things, from the fact that geographical forms are durable. The technologies and techniques that geographical forms embody and to which they give body therefore seem to have a finality linked to the preceding mode of production or one of its moments. Space as a form, therefore, is not phantasmagoric. Spatial objects are periodically rejuvenated by the transformation of society.

It can be said of forms in general that they change when their content or origin changes. Spatial form is different: it can incorporate *another* new form, it can adapt it or else destroy and substitute it completely. But then it would no longer be the same form.

Spatial forms are resistant to social change partly because they are material. The legal system is also resistant to change. According to Lukács (1960, 125), "the legal system should confront the individual events of social existence as something permanently established and exactly defined, i.e. as a rigid system." But its forms, though frozen, are not materially fixed like geographical forms.[6]

Space is never a finished product. It is never frozen forever. But one of its important elements is fixed to the ground.[7] Spatial forms, whether inherited or created, have the singular characteristic that, *as material form*, they do not have autonomy of behavior, but they do have autonomy of existence. This ensures that they have an original, particular way of entering into relation with other facts of social life. They have the properties of *thingness*. For Hegel, in *Sciences of Logic* (vol. I, Book II), "a thing has properties; these are, first, its determinate references to *something other;* the property is there only as a way of reciprocal relating.... But, *second,* in this positedness the thing is *in itself.*... A thing has the property to effect this or that in an other, and in this connection to express itself in some characteristic way."[8]

Now more than ever space appears as a solid unity. This is the basis of its specificity as an indivisible commodity. Infrastructures, by their very nature, are continuous.

In the unpublished chapter of *Capital* (French edition, 116), Marx wrote that railways and large constructions "present themselves as a *unique* commodity, as they do not accept metrical division." In other words, no *measure* can be validly applied to any of their sections. In the same way, within total urban space, it is impossible to evaluate an asphalt road, an unmetaled road, and an entirely undeveloped route in isolation from one another. All exist where they are found as local, internal manifestations of the uneven and combined development of society. Only total society can be the standard by which we appraise and value them.

Space as History and Structure

The debate over whether to analyze the real through the historical or the structural continues to be contentious. It is as much a philosophical as an epistemological question. The historical approach sees the progression of interaction and function moving from the past to the present. The structural approach addresses the relations between the variables that make up the structure of the contemporary situation. The opposition between two approaches, which take different routes to different outcomes, is a rich field of debate. But, when space is put into this field, we must conclude that spatial structure is the past in the present. Space works according to contemporary laws, but the past is *present*. Furthermore, present space is also future space. Built things and produced space are endowed with finality from the moment they occupy a place on the surface of the Earth. For Sartre:

> the praxis inscribed in the instrument by past labour defines behaviour apriori by sketching in its passive rigidity the outline of a sort of mechanical alterity which culminates in a division of labour. Precisely because matter mediates between men, men mediate between materialised praxes, and dispersal orders itself into a sort of quasi synthetic hierarchy reproducing the particular ordering imposed on materiality by past labour in the form of a human order. (2004, 184)

Through space, history becomes *structure* and form. Its forms, as modes of form-content, influence the course of history and participate in the global dialectic of society. The implicit or explicit question is whether space is at once support and factor. We can now give the first part of a response. Space is exclusively a support if, with George Edward Novack's (1969) irony, something can exist "in a given moment." But is there also something "outside the flux of time"? *To make,* that is, *to be a factor,* means being either the subject or the object of a process. The word *process* itself is another term for the passing of time.[9]

Like the other orderings of society, space is a social structure, but it has particular characteristics that make it different.

According to Lefebvre (1991, 73),

> (Social) space is not a thing among other things, nor a product among other products: rather, it subsumes things produced, and encompasses their interrelationships in their coexistence and simultaneity—their (relative) order and/or (relative) disorder. It is the outcome of a sequence and set of

operations, and thus cannot be reduced to the rank of a simple object. At the same time there is nothing imagined, unreal or "ideal" about it as compared, for example, with science, representations, ideas or dreams. Itself the outcome of past actions, social space is what permits fresh actions to occur, while suggesting others and prohibiting yet others.[10]

This is, after all, the *differentia specifica* of space. It is what gives it a particular role within the social system and ensures the (relative) autonomy of its development. It also names the specificity of the historical existence of space at a given moment. It is precisely what Kusmin (1974, 73) called "the specific logic of the specific thing."

PART III

For a Critical Geography

14

In Search of a Paradigm

S cientists have to take risks. None is more serious, though, than the risk of expressing scientific truth as if it were eternal certainty. Given the current mood of scientific work, this risk is particularly acute. Thinkers and researchers reach positions after lengthy reflections, but when these become dogma, all discussion ends up circling around the validity of a given hypothesis and the investigation of real things get sidelined.

This danger also applies in the (pretentiously termed) "exact" disciplines. Their object of concern permanently evolves, and new interpretations continuously emerge.

The expansion of knowledge is multilateral. Progress in one field is transmitted to others and affects them. Dialogue is fed by the proliferation of sources of information that can germinate knowledge in all directions. No science is immune. Inputs from neighboring laboratories mean received truths have to be reconsidered. As new truths solidify, each discipline has to modify, adjust, and improve its own understanding of reality.

All Theory Is Revolutionary

If we ignore the new and merely repeat old, totemic knowledges, we can fall into the trap Boulding (1969, 3) outlines: "rather than studying it, science takes upon itself the right to create the world that it is studying."[1]

This means renouncing science's core task of renewal. Theories are always, by definition, incomplete and vulnerable. If they are presented as if they have absolute value, doctrinal precepts become rigid obstacles to truth.

The potential of research is limited when it merely tests and verifies theories.

If science can create only what it already knows, it renounces its purpose.

The task of reformulation and reconstruction, therefore, is enormous. In geography, new theories are still bound to old ideas of space, as if the meaning of the basic parts of space had not changed. If we omit new elements and elide their precise meanings, it becomes difficult, or impossible, to construct adequate theories.

Theory has to fit the contemporary conditions of the world. It has to relate new things to old meanings and old things to new meanings.

In this way, all true theory is revolutionary theory.

Paradigm and Ideology

The idea of the paradigm has long concerned philosophers and scientists. Though Kuhn gets too much credit for the idea, his work drew attention to a number of formulations that have influenced mid-twentieth-century science.

The idea of the paradigm has had different meanings and criteria. In general terms, a paradigm is a framework for concepts, theories, and models. We often hear of paradigms expressing theoretical conceptions. The problem is to identify precisely how a new paradigm condemns an old one to oblivion and forces a renewal of the whole apparatus. This question cannot be resolved outside history. Observing facts in their concrete forms reveals new sets of systematically established relations to different specialists. Their capacity to displace previously powerful theories emerges from the fact that the new system of ideas is drawn from reality itself, and not from any particular philosophy.[2]

The validity of any philosophy is secondary to the proof of facts.

Antonio Christofoletti suggests that "in the development of the sciences, each phase is characterized by the dominance of a paradigm that expresses the theoretical conception which explains and orders scientific facts, guided by the need to formulate inquiry." He adds that "leading research goes on developing and approaching questions within the boundaries of the acknowledged paradigm, allowing problems to be approached which are no longer explicitly explained by traditional theory."[3] As new problems are presented, little by little a new theory emerges to incorporate them and resolve societal challenges. The new theory substitutes the old. It changes the sequencing and explanation of facts and alters the scale of value.

Christofoletti is right. Only now it is not the new theory that changes the "sequencing and explanation of facts, as well as the scale of value" (4) but the new sequencing of facts that encloses a new scale of values and forces the creation of a new theory. Contrary to Christofoletti, this is exactly why "quantitative" geography will never become a true paradigm: it does not interpret facts as they are but makes new facts according to a particular ideology.

An ideology is not a theory but its opposite.

Herein lies the risk of presenting an assortment of disorderly categories and promoting them to the status of theory. Such an approach cannot lead to a precise analysis of the totality; the whole of reality.

Nature as Paradigm

It is a century and a half since Ritter said that making theory is distinguishing a general system in nature. Nature here can be defined as the set of all existing things; reality in its totality.

Nature is in a state of permanent movement, and each of its moments is fleeting. Hence defining the present is always difficult.[4] To know the present is to discover the new conduct of beings in relation to one another.

For George Santayana (1924, x) the analysis of nature rests on "public experience. It needs, to prove it, only the stars, the seasons, the swarm of animals, the spectacle of birth and death, of cities and wars. My philosophy is justified ... by the facts before every man's eyes."[5]

In his classic *An Essay on Nature* (1940, 3–4), Frederick Woodbridge wrote that he was using the word "nature" as a name for "the *familiar* setting of human history ... the primary subject matter of all human inquiry."[6]

Finally, according to Whitehead (1964, 167–68), "the concrete facts of nature are events exhibiting a certain structure in their mutual relations and certain characters of their own. The aim of science is to express the relations between their characters in terms of the mutual structural relations between the events thus characterised."

Models of perception of reality change substantially whenever there is a profound technological, organizational, or social change.

We can't analyze the capitalist system as if it were the Middle Ages. African countries at the end of the nineteenth century require a different analytical methodology before and after their insertion into the modern capitalist economy. All continents are incomparably different after the end of the Second World War and demand a completely different set of explanations. A technological transformation is enough to change the social structure, and theory itself must therefore change.[7]

A new paradigm has to represent a complete change in the vision of the world.[8] Indeed, it is less our vision of the world that has changed than the world itself. Human history is marked by quantitative and qualitative leaps. They signify new combinations of technics and productive forces and, consequently, new frameworks for social relations.

With each technical change, old scientific truths cede their place to new.[9]

Let's not delude ourselves. We cannot think of a paradigm as limited to a singular, isolated science. A paradigm affects all scientific disciplines at the same time, not only the "exact" ones. If it is the case that with each new paradigm the relative importance of the sciences changes, this is not to say that some fields of knowledge escape the revolutionary impact of the new paradigm. It imposes itself on all the sciences in transformative and often brutal ways.

The problems to be addressed when a new paradigm emerges are not, therefore, *partial*. The whole problematic must be reconsidered, because the problematic of the whole is no longer the same. This does not mean that the totality of relations are all affected equally. If we want a valid interpretative apparatus, it is enough that some, or even one, of these relations experience a significant change (whether in terms of technics, modes of production, relations of production, or labor relations) for the whole theoretical edifice to crumble and have to be immediately substituted.

The idea of the paradigm cannot be drawn from the history of a single science or the happy discovery of one lucky scientist. It is historical; it emerges in synchrony with the background movements of history.

15

Total Space in Our Time

With the seminal exception of Henri Lefebvre (1973), all attempts to explain space have basically ignored the key problem of production. The practice of production is fundamental to making human knowledge. As Béla Fogarasi's work reminds us, "the most basic and abstract concepts have arisen in the context of labour-processes" (Schmidt 1971, 243).

It is in the parallel between the creation of the means of production, the productive process, and the production and transformation of space that geographical method emerges.

Production and Space

Nature was always man's storehouse, even in the presocial phase. But for the human animal to become the social human, he must become the center of nature. He achieves this through conscious use of the instruments of labor. In that moment, nature ceases to command the actions of men, and social activity begins to be a symbiosis between human labor and a nature ever more modified by it. This historical phase could not unfold without a minimum of social and spatial organization.

My approach is fundamentally based on recognizing that human space is, in every historical period, the result of production. The act of producing is the act of producing space. The human animal became the social human when he began to produce. To produce means to take from nature the necessary elements for the production of life. Production, therefore, supposes an intermediation between man and nature, mediated through the invention of techniques and instruments of work.

Man begins to produce when he works cooperatively with others—that is, in society—in order to achieve preconceived objectives. Production is the conscious use of instruments of labor with a defined, preestablished objective.

No production, however simple, can be done without the means of labor being available, without life in society, and without a division of labor. From this first social organization, man finds himself forever obliged to conduct a communal life, a "planned" and organized existence.

Thanks to its own forms and rhythms, production imposes forms and rhythms on the life and activity of men. These are daily, seasonal, and annual rhythms linked to the production necessary for the survival of the group. This new discipline enforces a rigorous use of time and space.

Such rhythms of life and activity create repetitive practices: hours dedicated to labor and hours dedicated to rest; the rhythms of production itself: the period of preparing the earth, the time of sowing, of clearing the fields, of harvest, of storing; the time dedicated to the communal labor of building, constructing houses and warehouses, building and fixing roads, and erecting infrastructures.

Each activity has its own place in time and space. This spatiotemporal order is not random; it emerges from the inherent needs of production. This explains why time and space are used differently in different historical periods and in different places. Their use changes with the types of production.

Thus, while *homo faber* becomes *homo sapiens,* a particular value is attributed to time, out of which emerges a specific organization of space: a particular arrangement of the objects through which man transforms Nature.

To produce and to produce space are indissociable. Through production man modifies First Nature—raw nature, natural nature—socializing what Teilhard de Chardin calls "the world ecosystem." Hence space is created as Second Nature—transformed nature, social or socialized nature. The act of producing is, at the same time, the act of producing space.

That which is created by life cannot be dead or immobile. As ways of producing change, the relations between man and nature change and the distribution of objects created by man to enable production and reproduce life change. If a new plant is domesticated and brought into production, it imposes a new order on time. This implies new localizations and a new organization of space. The animal incorporated into labor contributes to the modification of distance-time: another rhythm is imposed. The group acquires a new measure of time. When social time changes, space changes. These fields expand just as the fraction of time dedicated to rest, leisure, or celebration expands.

Eventually, a new technics is discovered. New techniques and technologies can be applied to labor, the preparation of the earth, storage, and the simplest acts of everyday life such as cooking. This is what we call, today, somewhat imprecisely, an increase in productivity. As the returns on labor increase, the time dedicated to work reduces.

Indeed, whenever the social use of time changes, so does the organization of space. All new technics are revolutionary when they reorder human space. Nikolaï Bukharin ([1926] 2011, 121) wrote that "technology is a varying quantity, and precisely its variations produce the changes in the relations between society and nature; technology therefore must constitute a point of departure in an analysis of social changes."

From one stage of production to another, from one way of ordering time to another, man is permanently writing his own history. This is both the history of productive labor and the history of space. Initially a history of an isolated group: a handful of men and a piece of nature *mediated* by the technics that the group itself devised to guarantee its survival.

At the dawn of social time there were as many ways of ordering time and nature as there were human groups on earth. Thousands of geographies existed at the beginning of history. But that time passed.

One contemporary challenge is to understand how changing human groups transform their relations with nature and thus alter history. The other is to configure the separate and multiple chains of cause and effect.

There are many causes of these transformations. We cannot exhaust the list or lay out a random classification. As a working hypothesis, we limit ourselves to considering a fact whose universality guarantees it general historical applicability.

Common, social labor defined by a shared objective and by a division of tasks that reduces the effort of each individual and the group, while also increasing its productivity, is called *cooperation*.

As cooperation increases, a larger section of areal space is needed for the group to realize its productive activity. For each smaller unit of time, the effects of each person's labor are greater.

As social production increases, the part belonging to each person theoretically increases. The existential minimum, however, is not very different between the richest man in the world and the poorest, so when a surplus exists, society's solution is to diversify production. For instance, more clothes are produced in the quest for improved living conditions. Artisanal activities emerge and develop. Intellectual labor—of priests and magistrates, teachers and artists, poets and seers—develops in parallel. The conditions for rest, creativity, and celebration expand.

New activities demand a place. They impose a new arrangement of things, a different disposition of geographical objects, a different organization of space from before.

When the phase of pure subsistence is passed, the surpluses of each group must be exchanged. But this primitive mode of commerce cannot itself change the structure of isolated groups. Since the goods produced and the form of producing them stay the same, the internal organization remains intact. So, therefore, does the ordering of time and space. The values placed by each group on time and space influence how they are organized.

In this moment trade is simple exchange or barter. Once trade becomes speculative, everything changes. In the stage of simple exchange, each side exchanges the same labor time. Even though it is represented by different quantities of goods, their value is negotiated, because neither group has the means to impose a fixed price.

Speculative trade introduces a new scale of values. The value of goods exchanged is no longer based on the quantity of work done for their production. From now on, this value is arbitrarily fixed, and the trader can only tweak or adjust its value in an equally arbitrary way. In this process, the product becomes a commodity.

The *commodity* is introduced into the life of a social group with the creation of a new social relation, *money,* the *cash nexus*.[1] It is a strange social form, even now, but it appears in the social group as a way of obtaining money to buy what is needed.

The old equilibrium is broken.

To buy commodities with money, you have to produce things that enable you to obtain more money, and so you must neglect what is less monetizable. The value of goods produced by the group is no longer furnished by their traditional role in collective life. From now on, the value of each product is given by the value, extraneous to the group itself, of the commodities they need to buy.

If the price of the commodities to be bought increases, new transformations are imposed on the group's mode of life. Fertile land is enclosed. Speculative trade separates out those who produce goods with speculative "value" and divides those who can buy commodities from outside from those who cannot.

From this moment we can speak of social classes and of differences in purchasing power. In this moment a real revolution in social relations begins.

From this moment, a new movement animates local *society*. (We can no longer speak of the social group that we had initially defined.) This movement comes from inside local society but also from the energies brought from other societies with which commodities are traded.

Time and space are reorganized and transformed as local society adapts to new productive processes and the new conditions of cooperation. Each renewal of the technics of production, transport, commercialization, transmission of ideas, ideologies, and hierarchies corresponds to a new form of cooperation which is deeper and more spatially extended.

The end of the fifteenth century saw huge progress in navigation, the militarization of the seas, and the introduction of trade and colonization in the recently discovered Americas. It was a key period in the transformation of the world's inhabited lands. The next fundamental transformation came at the end of the nineteenth century, with the creation of empires, the railway, the steamship, the wireless telegraph, and the banking revolution. These changed ideas of distance and scales of time and space. After these key moments in the history of humanity, we have arrived in the contemporary epoch of the scientific-technological revolution.

Networks of influence operating simultaneously across a multiplicity of scales, from the local to the global, transform space. In the end we have reached a world in which, more than in any other historical period, we can speak of total space.[2] Total space is globally imbricated space.

Total space and local space are aspects of a single and coherent reality—total reality—which is the image of the *universal* and of the *particular*. Global society and global space are transformed by time, in a movement that, though equally connected to diverse fractions of society and space, results from the interaction between global society, global space, and its diverse fractions.

The Universalization of the Economy and of Space

At the dawn of time, human groups drew from the space that surrounded them—from their bit of nature—the essential resources they needed for their survival. As the division of labor deepens, an ever greater part of the needs of each group, each community, have to be sought in the geographical area of another collectivity.

The idea of space as the biological support of human groups and their activities comes from Claval (1970, 11). It now needs a less literal interpretation. We can no longer apply it directly as the area of activity necessary for the existence not of isolated groups but of humanity in general has expanded. In our historical period, global space imposes itself through variables with ever more distant origins and global reach. This has become possible because, as the organization of contemporary capitalism is directed by multinationals, the process of capital accumulation is bound up with big companies seeking the conditions for the greatest profits wherever they may be. The expansion of raw and intermediate products necessary for the production of goods and the differences in the cost of labor between countries have deepened in the past thirty years. Meanwhile, a series of factors has contributed to the rapid globalization of the economy, including the expansion of transport and communications and a relative reduction in their cost (as a portion of the total cost of production), as well as the reduction or suppression of barriers to trade.

This process, begun with the globalization of consumption, leads, in the end, to the internationalization of production. Human groups, regions, and countries all consume growing percentages (in number and quantity) of goods originating beyond their own frontiers.

Both large and medium-size companies work in a global context.

This change reaches into our homes. Our everyday activities of biological and social life rely on products whose origins we cannot establish.

Everything that surrounds us carries the sign of a consuming internationalization. Even our bodies, imbricated with what surrounds them, do not escape this globalization. Our clothes, our shoes, and so many other things we use every day are produced not where we live and where we come from but thousands of miles away.

None of this would have happened if it were not for diverse, parallel processes of internationalization: of capital, of technology, of the market of goods and the market of labor, of education, preferences, and tastes, including for food. Alex Inkeles (1975, 467) points out the failure of perception of many sociologists and economists when it comes to problems of social change at the global scale.[3]

The idea of the ecumene—attributed to Sorre but with a longer history going back to Strabo—will have to be reconsidered. Demangeon (1943) guessed at this evolution when, listing the challenges for human geography, he included the valuation of resources and their changing social use, human distribution (considered as a reaction to natural conditions), and the expansion of human population in terms of efficiency, density, fluctuation, and migration.[4]

Still, the idea of the spatial distribution of humanity in relation to natural conditions is insufficient. The *habitat* of men was formerly the place where they lived and worked and the space of the geographically confined social life generated through both the material and nonmaterial aspects of the productive process.

Today, the space of societies is not merely the sum of each society's space. Neither is social space exclusively the *habitat* of men, thanks to the new nature of intrasocial relations and the relations between societies. The idea of space has become very different, and perhaps distant, from the idea of the ecumene. As new relations extend beyond isolated communities and countries to encompass the global, social space is much more than the set of *habitats*. The construction of space in our time results not only from immediate economic activity but also from economic speculation on the value of currently uninhabited or unvalued areas. A technological example shows this duality. Extraterrestrial navigation and new technologies of remote sensing enable countries to uncover their own previously unsuspected natural wealth. These technologies demonstrate how aspects of these countries evolved, but more significant is the concentration of scientific and technological knowledge in only two states: the United States and the Soviet Union. In fact, these two countries alone really know what resources the rest of the world possesses. This endows them with the conditions to dispute global power on scientific premises. As these two countries become more efficient at imposing their hegemony, there is a reaction. Other countries see that, now more than ever, they must defend their natural resources in a period when only one thing is certain: the material base of production is ever more limited. The result is a marked, and for some a surprising, volte-face by resource-rich countries in relation to their purchasers. The content of the idea of national sovereignty changes because even the poorest states, not knowing exactly what they have to defend, find themselves forced into a rigid ordering of the totality of their territory and its potential. They are *forced to defend everything.*

Paradoxically, such uncertainty can be positive. International politics has emerged as a tool for the transformation of national spaces. This process proceeds not in spite of states but through their intermediation. It relates not only to present concerns—currently there are few countries that can fully exploit their resources—but to the future need to maintain territory intact. This conditions transformations in the organization of space.

The idea of military and economic security—one of the dominant doctrines of international relations of our times—leads to the construction of roads, bridges, outposts, and artificial cities. It dislocates populations to strategic areas. The production of space ceases to be strictly a consequence of production. Political facts are key drivers. In the end, of course, the economy cannot be ignored—once a new sociopolitical space is created, the relations between men and transformed nature are a productive fact. If space becomes total through the universalization of production, we must nevertheless remember that totalization works from the most universal to the most local of levels.

Perverse Universalization and the Role of Internal Structure

A *perverse universalization* does not reach all actors and is not equally used by all agents. It benefits only a few and is to the detriment of many. The contemporary instruments of universalization, which are said to eliminate time, shrink space, and bring people closer together, in fact only conjure this miracle for some. How many really benefit from the ease of communication established at the global scale by the airplane or the telephone? How many, equally, can access universal, multiple knowledge? The roads that expand across countries and cities are used by only some. The ability to use apparently universal means of communication is in fact directly related to the power of each actor: the state, the company, or the individual.

This is, therefore, a perverse universalization, because under its façade of generalization is discrimination: the increasing wealth and power of some and the increasing poverty and vulnerability of the immense majority.

The Totality and Dialectic of Space

The idea of totality, taken in itself, has always been open to abstraction and confusion, unless it is allied to the concomitant idea of the division of the totality. The distortion of the notion of universality accompanies the possibility of the distortion of the idea of the totality. We need to remake our analytical apparatus and free ourselves from dogmatic methodology.

The new form of totalization corresponds to the era of technology and the multinationals. It demands that the national framework be taken as a viable scale of this totality and places a particular emphasis on the value of the internal, concrete structure of each country. It is through this concrete internal structure that so-called global values are expressed at the level of each social class, place, and citizen. This is what really counts.

The universal totalization of the present—the present mode of production—cannot realize itself (materialize and objectify itself) except through another totalization, mediated by the concept of the economic and social formation. In the contemporary moment, when nations aspire to be states, the social formation is confused with the nation-state. In truth, no other category is more useful for the study of space, because this category ensures that we remain linked to concrete reality.

Hegel equated the notion of reality with that of the dialectic. The idea of the dialectic overcomes the risk of metaphysical elucubration when we analyze the reality of space. The idea of a dialectical space in motion was clearly expressed by Spinoza. He defined the parallel notions of *natura naturans* and *natura naturata*. As Lucien Karpik (1967, 53) notes, these terms were used by classical German philosophy as a central category in a polemical distinction between dialectics and metaphysics. *Natura naturans* is nature as it is in continuous time, Time 1; *natura naturata* is nature as it presents itself in sequential time, Time 2.[5]

The concept *natura naturata* represents a reality that cannot be conceived as an idea or realized in fact without the conditions offered by the other reality of *natura naturans*. This reality is genetically primary. It is mobile, destined to inexorably transform into *natura naturata*. There is always a first nature ready to transform into a second; one depends on the other, because second nature does not come into being without the conditions of first nature and first nature is always incomplete without second nature coming into being. This is the beginning of the dialectic of space.

Space and Instruments of Labor

Nowadays, there are few areas on earth that can still be considered remnants of natural, raw nature. What appears to us as nature is no longer first nature but second nature: wild nature modified by the labor of man. This is easy to see in a city or an agricultural area. It is less obvious in areas where the modifications made by man are less visible.

Nature is transformed by production, and there is no production without the instruments of labor. From the beginning of historical time, the human-producer conceived of his own instruments of labor and built them with his own hands. He transported them every day from home to work and used them as immediate extensions of his body. There was an almost total communion between the human and the instruments used and manipulated in everyday tasks of production. Thus they made their mark on nature and transformed it.

As the productive process grew more complex and as the speculative exchange of the surpluses of production imposed new conditions, the instruments of labor became larger and more complicated. No longer were they appendixes of the human body carried daily in human hands but appendixes of nature. We can speak now of fixed instruments of labor. We can include, on the one hand, the means of production applied directly to production itself (a mill, a factory) and those related to other moments of production such as the circulation of people and products (vehicles, streets, bridges).

Evidently, we have to include how mechanical and, later, kinetic energy have substituted for humans. This same evolution means we have to consider all the instruments that have been created and perfected for the transmission of messages, ideas, and commands.

The world of created things is made up of ever larger and more fixed objects. The skeleton of space forged in the productive process becomes ever more rigid. In the evolution of the instruments of labor it is a long journey from the hoe to the city. It is a transformation that is as much qualitative as quantitative.

In spite of the fact that people, ideas, and products are becoming progressively more mobile, there is a tendency toward the increasing significance of immobile resources. There is also a tendency toward the specialization of the instruments of labor. At first, instruments of labor were polyvalent; today they are devoted to exclusive functions. This evolution needs to be presented in historical terms. It works differently in all countries and varies even within each country. Each country corresponds to a constellation of created resources and a particular proportion of fixed resources, related to a certain level of the forces and relations of production.

But the quantitative and qualitative distribution of assets varies within each country. This variation is one of the components of spatial differentiation and regional inequality.

The increase in volume, fixity, and specialization of instruments of labor contributes to endowing each part of space with a potential functionality that contributes ever more toward spatial inequality.

Much as in global society today, isolated social groups created, through the intermediation of production, a second nature, a geographic space. But, for the isolated social group, their social space was "their" geographic space, created by "their" production, "their" instruments of labor. Therefore, it was easy to analyze the processes by which each group created a space through the productive process. For this bygone isolated social group, analysis was easy because the scale of the variables was the same as the space occupied.

With the geographical extension of the division of labor now cover-
ing the earth, two phenomena emerged progressively and in parallel:
(1) The instruments of labor that used to be mobile have become ever
larger, more fixed, and more lasting. They are now designed to last for
hundreds and even thousands of years; (2) Without eliminating other
levels of cooperation, the division of labor grows at the global scale and
transformations are driven at different levels. The scale of variables to
be analyzed together is no longer exclusively the scale of the place or
space that directly concerns the social group but the scale of the place
and the scale of the world, the country, and the regions in which the
place is inserted.

Forty years ago, Bowman could argue (1934, 115) that man was capable
of choosing how the potential of his piece of nature could be realized.[6]
These times have passed. Nowadays, though adapted to preexisting conditions, a single model reigns everywhere.

Today the instruments of labor and of power and ideas for how to use
space all come from outside.

Distribution of Total Society in Space

If we approach things vertically, from the perspective of the historical
evolution of the instruments of labor, we find that the nature of social
space has changed, especially in the twentieth century. If we prefer a horizontal approach, starting from within each space, we find an irregular
distribution of advanced instruments of labor and the coexistence of old
and new instruments. Instruments of labor in one place are rarely of a
homogenous age. Places are internally differentiated by the degree of the
modernization of resources and by how these different kinds of resources
are combined.

As resources are capital represented by the means of labor and by
human labor itself, the social composition of each place is different.
Today's labor is done as a result of yesterday's labor. Living labor is, to
a certain extent, conditioned by dead labor. But, because spatial objects
do not evolve at the same rhythm, different gradations of the old and
the new are harnessed by living labor. Dead labor is the different social
and spatial forms that condition the objective realization of society as a
whole. If we take Brazil as an example, the use of instruments of labor in

cities like Salvador and Recife are not a direct, mechanical result of the functional reality of Salvador or Recife but a consequence of the overall needs of Brazilian society in its local manifestations.

The driving force is the social totality that constitutes a dynamic adaptation to preexisting conditions. It does so through a variety of political, economic, cultural, and ideological processes. The totality is the driving force. Each of these processes is itself a force that is altered when social reality moves through it to give function to geographical forms. In this particular sense, the process comes to an end. But in more general terms it is infinite for two reasons. First, society is in the continuous and permanent motion of its own transformation and the transformation of its processes. Each process succeeds another, whose characteristics are different. Second, it is infinite because, once the process is extinguished in the object to which it gave a function, that object becomes an element of society. Inanimate material is enriched by this living element of society, capable of generating other processes. Everything, though, begins in social reality. As Lucien Sebag put it (1972, 62): "the primacy of being comes from the fact that it is never finished, and this inconclusiveness is resolved in time." If we depart from social reality, it is only to return to it.

Structure, Process, Function, Form

The *being* is the total society. Time is *process*. Functions, like forms, are *existence*. The fundamental categories of the study of space are, therefore, totality and time. However, because spatial happenings are not homogenous, the ideas of place, area, and scale—a fraction of space within total space—are all important. What happens in one place is not independent of what happens in another, precisely because what happens is a product of the movement of the social totality. Curiously, it has been sociology and not geography that has best analyzed this. C. Moya (1970, 178) writes that "space is defined as the totality of positional relations that organizes the totality of actors." Temporal change is defined as the functioning of this structure and as a dynamic internal to the social system. It is a way of realizing the institutional expectations that configure the totality of functions connected to a position in space—that is, to a place. "In this way," says Moya, "the category of function unites the categories of statistical structures and their expression, dynamic elements and systemic variables."

The ideas of totality, scale, system, and time are mutually imbricated categories. In the same way that the definition of each one of these categories is impossible without the intervention of others, any analysis that does not take *all* of them into consideration *at the same time* cannot embrace total reality. Any analysis made without these concerns in mind will yield a false interpretation. The totality implies the notion of time because reality is both a state and an ongoing totalization. It is a situation undergoing change.

The idea of totality is intrinsically linked to structure, without which it is blind and confused. As the totality is the social totality, its corresponding structures are social structures.

The spatial totality, one of the structures of society, must also be dealt with in terms of underlying structures (a substructure of society as a whole and a structure of the spatial totality). Here we can turn to the question not only of places but also of subspaces, areas that have in traditional geographical language been referred to as *regions*.

Because social happenings—*geographical* happenings—depend on society as a whole, each particular event is a determination of that whole society. It is also a place of its own, defined according to an original social dimension that is at once temporal and spatial. Places, as well as areas, regions, and subspaces, are therefore only functional areas, whose true scale depends on processes.

16

State and Space
The Nation-State as a Geographical Unit of Study

With the emergence of world society, space has become total. The brutal trajectory of this unification, which began with the opening of the capitalist period around the end of the sixteenth century, has also been a trajectory toward diversification. It has secured the principles of unity and diversity in history. This trajectory is reaching its apex in this technological period, in which each nation is able to find its destiny only in the form of a state.

The New Functions of the State

The traditional notion of the state has dwindled in the political-economic conditions of our current technological period. Command of the world economy works at a global scale. International politics is founded in short- and long-term economic interests. Most countries are ignorant of their own national wealth. Within nations, the role of minorities is growing, and, provoked by the contradictory conditions of the current system, so is dissatisfaction among many populations, principally among the poor. This all contributes to withdrawing from the state a significant share of its functions and power. It also renders the state an indispensable instrument.

The contemporary roles of the state emerge from the new realities of the international economy. The unequal diffusion of technology has provoked considerable, sometimes extreme, differences in the prices of industrial products in different countries. Differences in purchasing power have deepened.[1] The movement of capital has reached an unprecedented level.

Exchanges have multiplied, while borders have been reinforced to coun-
terbalance inequalities at the international level.

Great movements of people that cannot be controlled or even guided or
oriented by the state are characteristic of the contemporary world. These
multiplicatory exchanges, commonplace in the international economy,
give the state an irreplaceable role. Even the big companies rely on the
state to defend their interests. The role of the state constantly expands
as it meddles more and more in areas previously reserved to the private
sector. The state is necessary to the system for other reasons, too:

(a) It is the entity most responsible for the penetration of innovation
    and the creation of the conditions of success for investment. As an
    instrument of homogenization of space and the establishment of
    infrastructure, it facilitates the success of invested capital, above all
    big capital.
(b) Through its own investments, the state helps attribute the greatest
    benefits and the smallest risks to big capital. This division of
    activities at the international scale ensures the continuity and
    reproduction of unequal divisions of wealth.
(c) Finally, in order to be able to prosecute these functions, the state
    assumes a mystifying role as a propagator of unrealistic ideologies
    of modernization, social peace, and false hope.

In this world of sharp contradictions, the proliferation of states is both
a necessity and a desire of imperialism in its current stage of expansion.
It therefore institutionalizes all kinds of state penetration. In an apparent
paradox, the nation crystallizes in the internal contradictions of technological capitalism, including in the growing aspirations encouraged by
marketing and diversification of consumption in areas from food to education. This paradox deepens because new kinds of consumption have to
be satisfied through the direct or indirect intermediation of the state in
one of its privileged fields, external trade.[2]

The state is the only possible intermediary between the mode of production at the international and national scales. It is up to the state to decide
how to open up to innovations, capital, and people. It is thus responsible

for complicity in, or resistance to, the interests of the global capitalist system.

This system turns largely on the work of transnational companies. For a long time these have been thought of as "multinationals," but a closer analysis shows that they rely on states to exert their power. States support them not merely to align themselves with powerful forces but also to strengthen their own economies in a phase in which even minor setbacks can have catastrophic effects on the development of the capitalist economy. In fact, it has already been pointed out (by P. Emmanuel, for example) that developed countries rely on the unbounded exploitation of poor countries to respond to the demands of their own workers. In this way, large transnational companies increase their power and grow their investment capital.

The complexity of their external relations and the needs of their local societies make nation-states socioeconomic formations *par excellence.*

Given that the internal structure of each country or nation is specific and is consolidated by domestic and international modes of exchange, the differentiation between nation-states becomes clearer. The global force of technological imperialism becomes ever more fixed and individuated.

As social forms and established legal totalities, *nation-states are the unit of geographical study.*[3]

The State in Underdeveloped Countries

In underdeveloped countries, modernization coincides with the expansion of state functions.

The new conditions of national society and economic dependence on developed countries means the state has to respond to an ever greater breadth of activities.

Economic dependency functions at the scale of national economic life. Its oppressive qualities are often difficult to perceive at smaller scales. Yet, whatever the extent of its external dependency, these lower scales necessarily also concern the national state. Moreover, at a certain level, a state becomes internally interdependent. This has consequences for the planning and organization of space and the relative importance of each section of territory. An externally oriented economy and national aspirations are both affected by the normative forces of long-term legislation as well as conjunctural decision making. Either can act as openings toward or brakes upon imported modernization.

In the underdeveloped world the state is ever more necessary thanks to the growing contradictions of external relations. These emerge from the crisis of the world system as well as from frequent crises in systems of internal relations inherited from previous periods. Contradictions in countries like India, Brazil, and Mexico, not to mention Argentina, Iran, and Turkey, result from a growth model that is inadequate. Elsewhere, the incapacity to follow such a model or to substitute it for another generates similar contradictions.

In the new edition of his classic book on underdevelopment, Lacoste wrote that:

> If we want to define "underdevelopment" such that we reach a better understanding of the world, of what happens or could happen in it, such a definition must be constructed for each State. It is also essentially within the framework of the State that the privileged capitalist minorities can be beaten, one after another, by each national or popular movement. (1976, 242)

The State as Intermediary between External and Internal Forces

We have seen that the current organization of the international economy confronts countries with new realities that transcend individuals and firms. Companies and people call for a higher entity capable of controlling disparate forces such as variable prices, the extreme speed of technological and administrative evolution, and the requirements of extended forms of exchange and human migration. It is no coincidence that the generalized globalization of the economy since the end of the nineteenth century has coincided with the creation of the modern state and its evolution into its current form. In fact, any cross-border individual or commercial action is effective only when mediated by the state.

The state mediates between external forces and the local spaces that transmit them. But it is not passive. In accepting external vectors of influence, it deforms them and changes their meaning, direction, and nature.

The reorganization of a subspace under the influence of external forces is always contingent on the state.

At the scale of the individual, new economic conditions have created new human needs of all kinds, from the economic to the spiritual. Yet the enormous scale of technological instruments and the inequality of opportunity that modernization deepens mean that individuals cannot meet these needs themselves.

This is not to mention the increasing social complexity and anomie of urban life, which renders the role of the state ever more important. As a public authority, it intervenes in more and more diverse fields to establish, or reestablish, "social equilibrium" and to help citizens meet the demands of daily life such as health care, education, transport, work, and leisure.

The choice made by power to determine how collective needs are satisfied is part of spatial reorganization. Each investment choice made by the state—even unproductive investment—renders advantages to a given place and transforms the organization of space. Thus, whether from the point of view of international relations or the problems of everyday life of the most humble citizen, the state is a crucial factor in the making of space. It must, therefore, be a fundamental element in the study of space. That state actions are marked by contingencies and limitations does not change this.

State actions are contingent because international dynamics beyond the control of dominated countries force immediate changes in everyday life. This contingency means that the action of the state in each subspace is part of more or less conjunctural interventions.

The state's actions in relation to human groups are, therefore, contingent on its dependent status. It does not have the power to overcome the "roughnesses" that define each piece of territory. Each piece of territory is defined by a history and a specific arrangement of people, tools, and activities. No external or state action can ignore this roughness. The action of the state, or what the state transmits, can thus be analyzed according to two hypotheses:

(a) State action directly creates new roughnesses or contributes to reinforce and change the nature of already existing roughness;

(b) Or, state action has little purchase on the substance of flows and, if it
    does create new roughnesses, it does so a posteriori and indirectly.

Clearly the state is the primary factor for everything that concerns space,
but even in small subdivisions there are developments that (from a statistical perspective) *momentarily* escape the power of the state. Depending on their scales, groups of variables and parts of territory respond to
interactions of many kinds.

It is important to recognize what the dominant factor at each scale is.
Thanks to its nature, conception, organization, and functioning, at the
scale of the nation it is certainly the state. But at smaller scales? We need
to take into account the subspaces in which people's needs are met and
in which companies and administrations make their local presence felt.

We must examine the outcomes of state actions as emerging from the
demands of internal forces which orientate and convene power. Among
these forces we could cite population, demography, urbanization, the level
of industrialization, consumption and culture, and so on.

We must separate what depends on the will of the state from what
does not—what constitutes a conscious action of power from everything
that can be realized outside it. We must analyze actions in detail in order
to understand whether or not they are indifferent to power and economic
and political contingencies.

State Action in Subspaces

There are three principal modalities of state action:

(1) Intervening to satisfy local needs. Its response can vary in quality,
    quantity, volume, or form, not always corresponding to the local
    scale;
(2) Intervening to satisfy regional needs at a particular point within that
    regional space;
(3) Satisfying national needs in local space, such as roads, taxes, fiscal
    exemptions, customs, trade policies, and protectionism.

Diverse subspaces are influenced by local, regional, national, and even
international factors.

Local influences are considered, here, at the scale of agricultural and urban units, characteristics of population, and characteristics of the models of localization of activities and of people.

Regional stimuli depend on the "vertical" economic competitiveness of different activities, the evolution of existing activities, or the creation of new ones. These stimuli can exercise a spatial or horizontal competitiveness, because each activity has a spatial scale. Changes in their relative importance affect the success of activities, constitute the conditions for their permanence, or cause them to crumble or disappear. The model of localization of activities and people therefore changes.

National stimuli are much more numerous. At the top of the list are the demands or needs of an industrial or urban zone, along with economic movements such as inflation and deflation, whose repercussions can be positive or fatal for a subspace.

International stimuli can range from questions of demand—quantity and quality, price, types of products (raw, secondary, or finished; mineral, agricultural, manufactured)—to changes in technology, price, and structures of consumption.

Under the current conditions of underdeveloped capitalist countries, international and local stimuli appear to have the most autonomy to influence the evolution of subspaces. The state, though, has a fundamental mediating role.

Phenomena such as measures against inflation or deflation give the impression that the state manages the independent variables of the transformation of subspaces. But these phenomena emerge much more from the vicissitudes of highly modernized regions. We can compare the significance of international stimuli with those from national industrialized regions, with the caveat that borders do not interfere. A dialectic exists between local and extralocal facts.

Even if the international is often dominant (and more so than it used to be), the state plays an ever more important role. Hence the dispute among the great powers over the political apparatuses of peripheral states. Extralocal actions seek to dominate subspaces in accordance with their own characteristics. The state adapts to both the imperatives of the producers of flows and to the receivers of flows. It adapts to subspaces and what forms them: population, companies, institutions. The historical and contemporary dependency of subspaces in relation to international demand endows them with a limited autonomy, exercised more forcefully in relation to the state than in relation to the foreigner. The state, meanwhile, remains the only organization that can oppose the various forms that external forces take.

The State and Spatial Transformations

In terms of the organization of space, the state depends on the powers it can reserve for itself.

A "primitive" state—basically a political organization distributing land among farmers—has decisive power that only rarely goes beyond the *local*. The regional implications of this depend on the regional economy—not only space in relation to one political organization but also neighboring spaces or those with which regional space is articulated. The state needs to function in a more sophisticated way if it is to control relations at a higher level.

The question is to know the level at which the state ceases to be a subspace and assumes a scale of decisions that influences a larger area. This is the level at which the state, by its own actions or presence, controls a more extensive network of relations.

If the scope of action of a political organization is limited to a subspace, it will not have the means to pass on the effects of its inputs and influences to other spaces. There is a hypothesis that the space-state constitutes the highest level of the structure or system that includes other systems or subsystems, though it is difficult to accept that it could be at once the top and bottom level of the structure. But this hypothesis is a dead end in a modern state, and all states are now modern, as nowhere has escaped the current imperatives of the modernizing historical system.

All state activities, above all in relation to the international economy, are marked by contingency. State action is fluid, constantly readapting to the conditions of international life. This contingency affects space and its reformulation.

The action of the state is marked by the need to take structural and contingent facts into account. Indeed, the conjuncture is as much that of the nation as of the region or the globe. The repercussions of state action on the reformulation of its internal space are marked as much by their contingent character as by the roughnesses of space itself.

Therefore, in analyzing the action of the state through the optics of subspaces, we must distinguish what directly creates new roughnesses. The state can create them or contribute to reinforce the character of those that already exist, even if this implies a change in the nature of the roughnesses in question. When the state decides to build a hospital, a school, a power plant, or an industrial town, it is creating a new roughness or contributing to reinforcing an existing one. The behavior of space depends as much on the actions of the past as those of the present.

Even though the state is the primary factor, some variables escape its influence. Models for how variables function differ because groups of variables or parts of territory are susceptible to different changes according to their respective scales of action. So it is important to recognize the dominant factor. On the one hand there is the scale of a country—the organization, the conception, and the functioning of the state—and on the other hand there is the smallest of scales—the spatial support needed for any activity to be realized in any given situation.

There is, therefore, always a dialectic between the macrospace (the state) and the microspace (collectivities of smaller dimensions). This dialectic is an essential subject in our analysis.

As a direct or indirect instrument for the local realization of external forces, the state also adapts itself to the exigencies of its own internal relations.

Space and Territory

Essentially, a nation-state is made up of three elements: (1) the territory; (2) a people; (3) sovereignty. The use of territory by the people creates space. The relations between the people and their space and the relations between diverse national territories are regulated by the function of sovereignty.

At its limits, of lines traced by common accord or by force, territory is immutable. This territory does not necessarily have the same extension through history, but at any given moment it represents a fixed point. It is called *space* once occupied successively by a people. This occupation results from the action and labor of a people, performed under the basic norms of a particular mode of production and the ever greater coercion of a sovereign power. It is the use of this power that determines the relations between social classes and the forms of occupation of the territory. We take up again here the argument developed earlier.

The action of territorial societies is conditioned within a given territory by (a) the dominant mode of production at the international scale, whatever its concrete combinations; (b) the political system, responsible for the particular forms that the impact of the mode of production takes; and (c) the impact of prior modes of production and the preceding moments of the current mode of production.

17

The Ideas of Totality and Social Formation and the Renovation of Geography

Some analytical categories are seen as permanent, while others have longer or shorter life spans. Some come to have particular relevance in particular historical moments thanks to unique conjunctions of circumstances.

We might imagine that the category of socioeconomic formation would be best used when countries have internal and external autonomy. Not so. In our own period, properly "internal" processes of production are externalized and "external" production is internalized. We have a deepening of dependency like never before. Furthermore, national socioeconomic structures have become more individualized and differentiated.

Just as the creation of colonies was a necessity at the end of the nineteenth century, the state has become a contemporary requirement for the capitalist system. The existence of a state attributes to the social formation a defined juridical, political, fiscal, financial, economic, and social framework. The economic and social structure of each country becomes ever more specific through its own transformations, which operate under the influence of many internal and external factors.

Multinationals have truly globalized the world. The idea and reality of a greater totality is being solidified, but it is totality lawless, except for the law of the strong. The globalized world is right under our eyes, as a fact, but it is hidden by the conditions in which it was created. When the economy is based on absurdity, social order is maintained only because

147

ideology intervenes. The political order ends up infinitely confounding itself with other parts of life. The "global" world presents itself to its observers, who are also its agents, as something beyond comprehension, at least for those individuals—the majority—who are the least well equipped.

However, the state, though more and more internationalized, appears to the naked eye as something fungible: a real framework capable of capturing and naming the determinations that define it, from its origin to now.

Dominated by one mode of production, the world creates objects according to a particular *historical order,* a history that involves the totality of all countries. It is through each social formation that a *spatial order* is permanently created and recreated. This spatial order is parallel to the economic order, the social order, and the political order and attributes a particular value of its own to the things, men, and actions that emerge from it. The social formation, therefore, constitutes a legitimate explanatory tool for society and space, respectively.

Totality and Space

The idea of totality has gained new significance as the lived moment of the capitalist system. This has a certain irony, as reinterpretating the idea will elicit new insights not previously envisaged.

Without the idea of totality, how would we explain that certain states get richer every day and others get poorer every day? How would we explain that in spite of positive and even, in some cases, reassuring figures of economic growth, the number of poor people is constantly increasing? How would we explain that in rich countries, where the surplus extracted by superexploitation flows into every corner, the number of people without work and the number of the poor ceaselessly increases? This reality is forever shown by the production and trade statistics discussed at all levels and in the most diverse media. To truly see this reality we must see the totality and the interdependence between all its parts. The idea of totality, made evident in the contemporary conditions of capitalist evolution, no longer allows us to put a sticking plaster over the ghastly wounds of the inequalities and contradictions of the system itself.

All aspects of social life are important. None of them, in itself, has primacy over the others. This is a guarantee against dogmatic and rigid epistemology that fails to take into consideration the totality and its movement. This impedes us from considering elements of "total society" as if they had the same meaning across time. We have to consider them as they are, as *variables,* whose value at any moment is given by the new totality created by the movement of the preexisting social totality and its transformation. Here we have the idea of empirical time: the time of real history and its concrete beings, of men, things, and actions.

All of this explains why the study of social and economic formations constitutes the best point of departure for such a focus. As a theoretical category, SEFs exist only because of their concrete aspects. These allow us to account for the specificities of each society (its particular evolution, current situation, internal and external relations) taken as a historically determined reality and founded on a territorial base.

The use of this category is valuable for a number of reasons: it means we do not stumble into "spatialist" approaches, it helps us to evade methodological positions that fragment reality, and leads us to analyze space through a problematic that privileges the social totality. I agree entirely with Alejandro Rofman (1974b, 14), when he argues that "in order to interpret the behaviour of geoeconomic space it is necessary to understand beforehand the behaviour of the global society on which it feeds." This position is far from the dualist thesis that sees society acting *on* space as if it were independent *of* space and vice versa. For the same reasons, we must refuse all fetishizing interpretations that would attribute value to forms themselves. As J. L. Coraggio (1977, 23) put it, we are interested in spatial configurations only "in that (a), they express social relations and (b), constitute conditioning factors of relations between specific agents in a socio-economic formation." This set of premises help us see the whole problem of space as a set of fixed and mobile resources—the English word "assets" might be more appropriate—and, at the same time, as an element of society.

This analysis makes sense, according to Lefebvre (1991, 299), only if "political economy has been reinstated as the way to understand productive activity. But a new political economy must no longer concern itself with things in space, as did the now obsolete science that preceded it; rather, it will have to be a political economy of space (and of its production)." Lefebvre is referring to political economy as an exhausted discourse.

But this critique can also be applied to geography when it refuses to study the real or to consider society in the entirety of its historical movement. The recent evolution of our own discipline has served only to increase the pallor of an already faded reputation.

François Simiand, a sociologist of the school of Auguste Comte, vigorously criticized geographers for "writing monographs in which the interdependence of things, and their dependence on the whole, disappear."[1] This reproach from two generations ago continues to be valid.

This is not a question of studying the whole through the whole, in which the risk of tautology is ever present, and we can, of course, deal with one of its aspects more than others. It would be wrong to consider ground rent, or the form of the surplus, or the geographical expression of class struggle, or the ideological role of architecture and urbanism, as if each of these categories were not what they truly are: a moment, a "region" of total reality, with a simultaneously subordinate and autonomous structure, endowed with their own determinations. The total thing, as *The German Ideology* (Marx and Engels 1947, 28) states, "can be shown in its totality (and therefore, too, the reciprocal actions of these various sides on one another)."

Social Formation and Space

The corrective to methodological errors that lead us to perceive reality wrongly is to put the ideas of human space and social formation side by side. (I elaborated on this in a recent article [1977a,b,c,d].)

Will we, then, be doing geography? I, like Lefebvre, sometimes speak of spatiology. Many of us, however, would prefer to keep the old name: a debate over terminology does not lead us very far. What matters is to be conscious of what is at stake.[2]

I propose the objective of this renewed geography to be the study of human societies in their process of permanently reconstructing space inherited from preceding generations, across the many instances of production.

This renewed geography (spatiology?) will concern itself with human space as it is transformed by the parallel and interdependent movement of a *history* made at different levels, from the international and the national to the local. The ideas of totality and structure, the universal and the particular, must be united in a single transformation in which society will be recognized to be in dialogue with transformed nature, not only as an agent of transformation but also as one of the results of transformation. All dualist temptations will be exorcised.

A totalizing concept such as the socio-spatial formation is best equipped for a task of this kind. A detailed formulation of this concept is presented in my earlier study.[3]

"The principal problem in human geography," argues Nicole Mathieu (1974, 71), "is the identification and classification of social forms which emerge from the unequal development of the productive forces and transformations in social relations." Revitalized geographical science would be the discipline of socioeconomic-spatial—or just socio-spatial—formations. We can speak exclusively of social formations, as these do not exist outside space. Such a study would assimilate the history of production and the history of human space into a single history of global society.

Spatiology would thus be a quasi-history, what Croce (1968, 85) puts into the category of meta-historical sciences: "a classification of discrete things through space [my emphasis]; a *meta-history*."

The Idea of Social Formation

The concept of the social formation allows us to go beyond the almost cognate idea of social structure. It must not be confused with the tired idea of total society.

Total society is not the same as the social and economic formation. For Jean Bancal (1974, 224), the two categories are interchangeable because "a socio-economic system refers, first of all, to a global society. This is the situated, acting economic whole; a coherent, extant social totality, in tension and transformation, animated by agents and groups."[4]

Following Jacques Berque (1970, 152), for whom, like everything in life, "social reality is divided into parts and hierarchalized," Bancal outlines "an effective realization of society (its mode of life), a pragmatic regime (its practice of life), a theoretical model (its rule of life), and a movement of its own (its living dynamism)."

But the commonplace expression "global society" does not denote the same thing as the social formation. The dominant conception, says Godelier (1974, 32), is based on Talcott Parsons's model, in which society is seen as a global system that articulates economic, political, religious, and other subsystems, each of which has a *specialized function*. The idea of the totality is excluded and the idea of a social formation endowed with an internal structure and external relations is suppressed. It is a Spencerian set of ideas.

Valentina Gerratana's (1973, 62) formulation explains the distinction pithily: "the social system is any old form of society, the social formation is society itself." A society can be unambiguously analyzed only in terms of the determination of successive histories. Only thus will we be able to understand its specificity and to distinguish its *essential qualities* from those of other societies. Once we define society according to its successive historical determinations, we will be able to proceed in a fruitful analytical direction. "Society" cannot be a scientific term, writes Pierre-Philippe Rey (1973, 165), if it is merely "designated by the finger that points at it, or by the name given to it. It is impossible to define society if the number of criteria used to define it is equal to the number of its observers." A concern with objectivity must emerge from the objectivity of what is to be defined. It is only through the systematic reproduction of the history of production that we can attain and defend objectivity.

A socioeconomic formation is "a totality, a system defined based on its own making." It does not proceed "randomly, but by following laws which systematically express a given mode of production or the passage towards another mode of production" (Aguilar 1974, 93–96). For some authors, this idea becomes a general concept. For V. Kelle and M. Kovalson (1973, 41), the idea of the social formation "brings out the essentially general, typical aspects of the order in the various countries at one and the same stage of historical development, the general which lies beneath the individual specifics of their history." This argument is frequently aligned with formulations of a "Latin American" or "African" social formation.

It is certainly useful to consider similarities among countries on the same continent. But it is an exaggeration to deny the specific role of historical and present accumulation in the formation of the individual characteristics of each country.

Social Formation and National Reality

A clear distinction between the idea of the mode of production and the idea of the social formation is indispensable. The former interprets the value of all kinds of forms in society, including geographical forms, in temporal succession. The idea of social formation enables us to interpret the accumulation and superposition of forms, including the geographical landscape. I dedicated a significant part of an article in 1977 to this crucial discussion.

The category of the economic and social formation is thus extremely useful to the study of national reality as it is not applied to *society* in general but to *a particular society*. The specificities and particularisms of each society must be attended to so that concrete study can yield concrete action.

Politics has become profoundly relevant for international relations. Anouar Abdel-Malek (1977) has analyzed the dialectic of imperialism and its implications for geopolitics and the spread of national independence movements. The significance of politics for spatial change is growing. In a synthetic study, Sônia Barrios (1977) has shown the importance of state behavior to the remaking of space. The state became the largest creator of infrastructure, activity, and employment whose siting is not necessarily subordinated to the law of the market.

Social Formation and the Renovation of Geography

So it is not difficult to see the importance of the idea of the social formation for the practice and growth of geographical studies.[5]

Indeed, we might suggest that the idea's almost total absence in the available techniques of spatial interpretation is one of the reasons that geographical theory has developed so slowly and one of the explanations for its inability to construct concrete and epistemologically coherent foundations. The ideas of mode of production and social formation— above all the latter—have not been adequately explored by students of human space.

I have written elsewhere (Santos 1977a,b,c,d), and I remain convinced, that the inheritors of Marx made a great error when they developed the idea of social formation as a category of social analysis without taking space into account.[6] We should be dealing more with the category of socioeconomic and spatial formation, for there has never been a social formation independent of space. Society can't be seen objectively without geographical forms. The objects that make up the landscape orient the evolution of society itself. This fact has never been sufficiently or system-atically taken into account.

The historical study of the formation of space after the arrival of the capitalist mode of production will form the basis for such an approach, and the study of economic and social formations are the best founda-tions for an investigation of this nature. As a theoretical category, social formations do not exist except through their concrete aspects. Concrete modes of production allow us to understand the specificity of each soci-ety (its evolution; its internal and external relations; the composition of its technological, capital, and labor conditions) as a historically determined reality with a territorial basis.

Studying a social formation must account for two sets of definite rela-tions outlined long ago by Lenin: horizontal relations and vertical rela-tions. Horizontal relations are the internal structure of society; vertical relations are those between a society and other societies. In the final analy-sis, these two sets of relations are interdependent. Recognizing this means acknowledging that the evolution of any given country is of interest not only to itself but also to others.[7] The degree of interdependence depends on the level and quality of relations.

The study by Alejandro Rofman and L. A. Romero (1974) is an exem-plary application of the category of social formation to the concrete case of a depressed zone in an underdeveloped country. It correctly deploys analytical categories and, departing from a well-established empirical base, advances theory itself and provides a rich specimen of the meaning of the particular within the general. Based on a concrete situation, it shows how the fundamental multiplicity of particular situations does not con-tradict the unity of history.

George Coutsinas and Catherine Paix's (1977) study is an example of another approach that relates fundamental aspects of the mode of produc-tion to the reality of underdeveloped countries. Using a synthetic concept of international trade, after patiently elaborating laboriously gathered fig-ures, the authors collate families of countries and highlight their essen-tial characteristics.

Only the history of world society and the history of local society can serve as a basis for geography—or, more broadly, the sciences of space—to interpret human space as a historical fact. This is critical to understanding spatial reality and the efforts to transform it and put it to the service of man. History cannot be written outside space, and space itself is social. There is no a-spatial society.

The idea of the social formation as a category of reality and as a category of analysis constitutes the best means to formulate a valid spatial theory. It deals with the differential evolution of societies in their own frameworks and in relation to external motive forces. The basis of explanation is production: the work of men to transform, according to determined historical laws, the space that confronts the group.

This approach means avoiding characterizing the relations between man and nature dualistically. This has been key to the misinterpretation of reality in many social disciplines, most seriously in geography, centered as it is on interpreting the relations between society and space.

As soon as we consider nature as transformed or socialized nature—*second nature,* to use Marx's expression—nature and space are synonymous. Neither Marxists nor geographers—nor even Marxist geographers —have given this the attention it deserves.

My ambition is precisely to explore a new possibility for the interpretation of the fact of space. This is closely linked to what Barrios (1976, 1) has proposed: "a conception of space that exceeds the frontiers of ecology and opens onto the social problematic."

I propose not to extend a single part of the semantic debate over social formations but to add a new dimension to it—space.

I understand all the risks of such a project. There is a strong chance that I will fail to deal with innumerable elements. My ambition, however, is to raise problems and instigate debate.

As geography seeks new paths, I want to offer a theoretical framework that is both universally applicable and does not deform the individual realities of each country.

18

The Idea of Time in Geographical Studies

David Harvey writes (1967, 550) that "just as Marshall came to regard the spatial dimension as relatively unimportant to his economic system," in Isard's terms the "Anglo-Saxon preconception" (1956, 24) led geographers to neglect temporality. Sauer (1963, 352) blamed Hartshorne for this.

Whether in *historical geography* or the *retrospective geography* of historians, addressing the idea of time in geographical studies is not new. Neither area has moved beyond posing problems to offering reasonable solutions. On the other hand, as already outlined elsewhere in my work (for example in *Economia espacial: criticas e alternativas* [São Paulo: Hucitec, 1978]), the idea of the *diffusion of innovations* has not moved forward precisely because it lacks a conception of social time.

The concept of *relative space,* lauded in the 1950s and 1960s in opposition to the idea of *space as container,* proposes that we abandon tridimensional space inherited from Newton and move instead toward quadri-dimensional space. This has become thinkable since Einstein introduced new possibilities into physics and philosophy.

The Diffusion of Innovations

The theory of the diffusion of innovations sought to provide intermediation between space and time. But the dominant concern with building deductive models halted its progress. Warneryd's focus on spatiotemporal systems remained undeveloped, and this promising theory became just an appendage to the geography of trade.[1]

The title of Ronald Abler, John S. Adams, and Peter Gould's book on the spatial distribution of innovations, *Meshing Space and Time*, for instance, conjures the image of a marriage between forms and society in motion, intermediated by waves of diffusion. But that's not how things actually happen. The same innovation can embed itself here one day and elsewhere tomorrow. It can reach one person today and another, far away, tomorrow. The theory implies that diffusion is a snowball effect. This suggests an unrealistic equality of conditions across places and people and a geometric progression or regression that affects places and people in a rigid temporal order. These famous waves of diffusion do not, in fact, exist.

Still, in spite of its weaknesses, the diffusionist theory has formulated and renewed spatial and planning theories based on "contagion" or hierarchical filtering down: theories of central places, growth poles, center–periphery, growth, and urban and regional planning. These do not take us very far because they lack a basis in objective reality. They lack the adequate theoretical orientation or methodological rigor to associate time and space within analysis of spatial organization.

To begin with the strictly temporal dimension of quadri-dimensional space, we must first define the four dimensions of human space. We must also consider the formative elements of space as analytical or methodological categories: that is, as epistemological categories.

The Spatial-Temporal Approach and Empirical Time

By existing, everything articulates the present, the past, and the future. So neither a merely spatial approach nor a merely temporal approach is sufficient. As V. A. Anuchin (1973, 52) wrote, the logic of time unites time and space, the two aspects of the existence of matter. The idea of space-time as a category of geographical analysis, however, has been the object of endless confusion. In its old form it cannot move spatial analysis forward.[2]

Quadri-dimensional space is promising because it reinforces the idea of relative space as a system of relations or as a field of forces in which time is an essential dimension. This supposes that time can be defined in properly geographical and not geometric terms and that it can be considered objectively and not subjectively. Perception, if taken as an exclusive attribute of the subject, cannot aid theoretical construction.

Conceptually, time must be measurable for it to be defined as a geographic variable. But caution is required. Measurement means not rigid and unchanging quantification but empirical extension. The problem is more of extension—the properly spatial—than of metrics as such. The contemporary occurs in space (Weiss 1958, 1–21) and "empirical meanings emerge spatiotemporally" (Moya 1970, 155).[3]

The fact that events are at once spatial and temporal does not mean we can interpret them outside their own determinations or without accounting for the totality from which they emanate and that they reproduce. Social space cannot be explained by social time.

This approach would be inadmissible if time and space were simply a kind of intuition, as proposed by Immanuel Kant,[4] Henri Bergson,[5] and Samuel Alexander.[6]

As G. W. Leibniz reiterates, space and time are not simple relations between objects and events. Only a Newtonian approach to absolute space and time—as opposed to time as the object of experience—could see them as such.

The analysis of the production of space, therefore, must begin from two premises:

(a) Time is a relative, not an absolute, concept. It is not the result of individual perception but is concrete. It is differentiated, with sections endowed with particular characteristics. The consequent periodization is based in empirical parameters considered not hermetically but as sets of relations. With this approach we can identify temporal systems.

(b) The relations between historical periods and spatial organization must be analyzed. This reveals a succession of spatial systems in which the relative value of each place changes with the movement of history.

The Need for Periodization

The idea of time is inseparable from the idea of system. At each moment of local, regional, national, or global history, the action of variables depends on the general conditions of their situation.

Using historical realities to explain the present does not always mean that the idea of time has been correctly incorporated into the study of space. The part of space under analysis must be understood within its temporal system if we are to argue that the study has a spatiotemporal approach. Simply referring to the historical backdrop of a phenomenon or looking for partial explanations (for one or other element of a whole) is not enough.

Most regional studies have precisely this flaw. It is of limited use to relate the present situation of a variable to its past status. It curtails the significance of the variable across time. From a geographical perspective, it is the succession of systems that matters, not simply variables or isolated subsystems. Space is defined by an integrated combination of variables and not by one or some of them, however significant they may be. Each variable is entirely meaningless outside the system it belongs to. Isolated variables become less hermetic through local processes of interaction. The making and remaking of spaces (their formulation and development) are chemical processes: the individuality of the space is the outcome of its particular combination. The continuity of space emerges from the fact that each combination is a function of the combination that preceded it.[7]

Just as individual elements cannot evolve without dragging the rest along with them, our problem is not the particular evolution of one element but global evolution.

Variables generally change quantitatively during processes of transformation—they conform to what Klir (1966, 30) calls the *activity of the system*. The meaning of transformation varies according to the scale—that is, the "*resolution level*"[8]—at which it is approached. Yet it is through the transformation of the ensemble, of the whole, that qualitative changes accrue real expression. Each variable ends up having a different meaning in relation to others.

The great lesson here is that in each historical period the value of each variable changes. Hobsbawm (in Marx and Engels 1964b, 17), is absolutely right that "economic development . . . cannot be discussed except in terms of particular historical epochs and particular social structures." If we forget this we risk losing ourselves in a history we do not know how to interpret. Moore called this temporal myopia. "History is a predicament for man who must live in it . . . the perspectives which he finds often merely reflect his age; and what he accepts as timelessly true and valid is apt to be merely the opinion which is in fashion," wrote Emil Fackenheim in his book *Metaphysics and Historicity* (1961, 1).

The reconstruction of successive temporal and spatial systems is key if we are looking to explain contemporary situations. This requires us to precisely identify periodizations at different levels or scales, as well as to isolate (for methodological purposes) the dynamic factors of each period at each level or scale. In any given case, we may have to take the direct or indirect role of the accumulation of capital at the global scale into account, as well as its repercussions at various geographical scales: the country, the region and subregions, cities, and villages.

No element can be considered in isolation. None exists outside the relations of totalization. But this should not stop us from recognizing the hierarchy of variables in each place at each moment. If we cannot, we might as well give up the task of correctly interpreting and defining space.

Space as Unequal Accumulation of Times

Once realized in a part of space, a variable (of different types and ages) crystallizes as a new fact, a new quality endowed with the capacity to make new relations. Up to a certain point, these different combinations condition variables' encounters. Localizations are historically determined by new and old combinations of variables.[9]

Our challenge is to understand the mechanisms of the spatial transcription of temporal systems. If the impact of a system of time on a fraction of space were fixed, each temporal system would make its mark fully on each part of space it touched.

However, as the action of previous historical systems leaves residues, different systems leave traces, except in blank spaces touched for the first time by external modernizing forces.

Apart from these extreme cases, the same subspace was and is at any moment the sphere of action for contemporary systems working at different scales. The hierarchy of innovations corresponds to various levels, and modernization generates specialization: a rare, spatially selective variable can gain huge importance. Specialization causes polarization. The position of the pole ends up at the most modernized and most specialized subspace. Other subspaces receive impacts from multiple origins with diverse connotations. The subsystem that corresponds to a subspace is subject to the more or less complete and durable control of systems that operate at a higher degree of resolution and at a higher scale. Thus emerges the hierarchization of space.

One thing, though, is certain. As in each system there is a combination of variables working at different scales, and a combination of different "ages," each system diffuses elements with different temporal markers. The dynamics of reception of the subspace itself are selective: not all "modern" variables are incorporated, and not all the variables that are incorporated are the same age.

Thus the problem of sedimentation appears in full. It is not only a question of the sedimentation of temporalities, because at any given moment the elements that enter into combination are of different ages. There is the concurrent sedimentation of influences from multiple origins and multiple spaces. All of these give each place a specific combination, a particular meaning that is simultaneously temporal and spatial. We could speak of a *spatial time* proper to each place.

The Idea of "Spatial Time"

Space is characterized, among other things, by the age difference between the elements that make it up. This is valid for all types of subspaces, no matter the scale.

In agricultural space, cultures, routes, houses, techniques, instruments, and populations have been embedded in the landscape for many years.

In urban space, activities, houses, streets, avenues, and neighborhoods are not all of the same age.

Whether we increase the scale of our observations to the continent or restrict them to a tiny cell of territory, we will never find entirely synchronous elements. Each variable was installed at a different time. Each place is distinguished by the multiple temporal markings of its constituent variables. In each continent, country, region, or subspace, each place represents the sum of particular actions initially located in different periods. The simultaneous presence of variables with diverse ages is the result of the characteristic combination of each place. It is *unique.*

Place, then, results from multilateral actions realized at irregular times across all parts of the earth. A theory that explains particular locations must take the local and more-than-local actions of the present and the past into account. Place thus ensures the unity of continuity and discontinuity. This both enables its evolution and ensures its unmistakably concrete structure. From a genetic point of view, at a determined point in time, the variables of space are *asynchronic.* They encompass the differences of "age" at the pole and at the periphery. Nevertheless, variables function *synchronically* in each "place." They all work together, thanks to their functional relations. Each place is, at each moment, a spatial system, whatever the "age" of its elements and the order in which they installed themselves. Being total, space is also punctual.

Within each historical system, variables evolve asynchronically, but the geographical system changes synchronically.

One spatial system is substituted for another that recreates its internal coherence, even if each isolated variable has its own pace of change. Thus synchrony and asynchrony are not opposites, but are spatiotemporally complementary, because the variables are exactly the same.

Taking into account the lag between variables and their corresponding segments of time, we can explain the differences between the organization of space between countries and what have come to be called "regional inequalities."

The Roughnesses of Space

Let us return to our central theme. Human formations are defined by local combinations of variables, which give them their originality. Of these variables, some result from contemporary flows and others from old flows that have been transformed in place. At any given moment, however transitory, formations can be defined by variables that are already "old," whose evolution was, briefly, endogenous. When new flows arrive, bringing new or renewed variables, they are warped by those already present.

At any given moment, a place can, for some distinct period, be immune to the fluctuating influence of the variables of a new historical phase.

Broek is right to argue that "geographical distributions are not only determined by positions in a functional system, like the positions of rubies in a watch, but they are equally and principally the result of non-recurring historical processes."[10] Boulding, equally, reminds us that "if growth creates forms, forms limit growth, the relations between growth and form being the essential key to understanding structural growth."[11] Spaces—and the combination of structures that characterize them—are, at any given moment, more or less oppositional or open to new influences.

In each occupied or empty place there is, therefore, a specific receptivity to the flows of modernization and innovation.

Conclusion

Geography and the Future of Man

Any science worth its salt should concern itself with the future. A human science must be concerned with the future not as an academic exercise but in order to direct the future for man: *for all men* and not only for a small number of them. This is man as a project, not a human animal: man in control of natural forces that guarantee and reproduce his own existence.

Now, as nature modified by human labor is more and more hostile, those who study it must redouble their vigilance. Geography, so often in the service of oppression, must be urgently reformulated to become what it always desired to be: a science of man.

Market-Space and the Geography of Class

The construction of space is the work of society in its uninterrupted historical march. Merely recognizing that space is the result of the accumulated labor of global society is not enough. We must work with a concrete idea of class society.

Society is transformed in space through its redistribution across geographical forms. This is to the benefit of some and the detriment of the majority; it separates people, attributing to each his or her part of space according to commercial value. Market-space is available to consumers as a function of their ability to pay. The study of space demands that we recognize the agents of this process, the place that belongs to each of us, whether as an organizer of production, the owner of the means of production, or the supplier of labor.

We have reached the moment when a New Geography can be created. Man is beginning, little by little everywhere, to recognize in the space he has worked to create the cause of so many of the evils that afflict the world.[1] As the Uruguayan G. Wettstein (1973, 7) puts it, geographers now have two alternatives: "to justify the existing order by obscuring the true social relations of space, or to analyse these relations, their contradictions, and the possibility of destroying them."[2]

For a Liberated Geography

It is not hard to choose between a liberated geography and today's geography of the *gendarme*. Woodbridge (1940, v–vi) writes that he investigates nature not only "in the pursuit of knowledge, but in the pursuit of happiness ... because in a world wholly devoid of any desire for happiness, the advancement of knowledge would serve no purpose." The guiding light of this kind of geography must be that to understand reality is to understand how it is produced.

From a genetic point of view, space can be analyzed through the reconstitution of the history of its production. But the reproduction of space emerges from the class struggle forged in the productive process itself.

Only by studying the history of modes of production and social formations can we understand the true value of everything in the totality. The totality that is the object of our research is very different from the partial universality of a system in which the privileged, in order to impose themselves on humanity, must, from the outset, anesthetize it. This universality is not the truth that the philosopher hoped to find in nature. It is a distorted nature, mediated by a society that no longer knows where ideology begins and ends. The science and vision of the world that emerges is necessarily underpinned by epistemological alienation.

This explains why the spatial science we desire is not official geography. Geography, the "widow of space," is not the spatial science that it has to be. It develops and maintains "ideological knowledge" while other spatial disciplines supply the instruments, methods, and techniques to turn concrete realities into ideologies in the service of big capital.

The new knowledge of spaces must denounce the mystifications of the sciences of space.

This New Geography, committed to the social interest, must take into account that we live in an epoch of transition, in which the nation imposes itself everywhere. Seeing the contemporary historical phase as transitional, we must not get trapped in the present as if it were the eternal. We cannot content ourselves with mere structural analysis.[3] We must address change. While recognizing the nation as a fundamental reality, we must recognize the key role of internal relations.

Cause and Context

These priorities oblige us to think more in terms of context than cause. The cause is only one instance in a global movement. Sometimes it has little meaning for the movement's reality. Often, when considering cause and effect, we lose sight of the process by which we bridge the past and the future. We see only the field of the sensible and the partial. Only through context can we see the motion of the whole.

To consider cause and effect is to work with the visible. To consider context is to work with what we cannot see, which is often more important.[4] The invisible becomes fundamental to explanation. It leads beyond form and appearance to offer us that which is beyond the phenomenon. Only in this way can we separate the "real" from the "not real" and distinguish the ideological charge by which space is attributed value as a commodity. This preoccupation is crucial today, when space is man's home and his prison. To get past this contingency is a complex interpretative task. It can be done only if we isolate, within the total movement of reality, what is false and what is imposed as an addition to the necessary by vested interests.

The Wheat and the Chaff: The Separation of the Ideological

Production is a domain in which ideology imposes itself brutally as a necessity of the system's survival. Since the act of production is also the act of producing space, it emerges under the sign of ideology. The mercantile creation of space is in itself a speculative, deceitful game. The *marketing* of space makes the trick seem truth.

Picking up Kosik's concept, we are limited by "pseudoconcrete" facts. We have to ceaselessly push to separate the ideological from the nonideological. In the global movement of society there is no other way of discerning that which necessarily brings falsehood. This is key if we do not want to get stuck contemplating the present but think of the future as a set of possibilities—the kingdom not merely of repetition but of the truly new.

We must demystify space, in order to link it with the social formation through a theory grounded in the real. This requires confronting space as it is: a social structure, like other social structures endowed with autonomy within the whole, and participating in an interdependent, combined and uneven development.[5] But the weight given to space in the evolution of societies does not mean we can see it as an autonomous concept, to be studied separately and independent of science and society. There is a significant risk of taking the appearance for the essence, of privileging the landscape to the detriment of the overall structure that animates it. This would mean falling into the trap—identified by Marx in his 1841 discourse on nature (1973, 121)—of confusing reality with its malevolent appearance.[6]

To demystify space we must take two key sets of facts into account: on the one hand the landscape, the functionalization of the techno-productive structure, and the place of reification, and on the other hand total society, the social formation that animates space. Thus we can demystify space and man.

To demystify man and space is to extract from nature the symbols that hide its truth. It is "to make Nature meaningful, and make symbols natural" (Dorfles 1972).[7] It is to revalorize labor and revalorize man so that he is no long treated as mere exchange value. We have in front of us both a problem of knowledge and a problem of morality.

Space and Liberation

More than in any other historical era, in contemporary conditions space has played a determining role in the enslavement or liberation of men. "The relations between man and nature," writes Victor Ferkiss (1974, 102), "are the central political problem of our time because we cannot understand relations between one man and another without knowledge of the relations between man and nature."[8]

In the more than thirty years since the Second World War, a great number of geographers, consciously or otherwise, have collaborated perniciously in the expansion of capitalism and all forms of inequality and oppression in the Third World.

We have to equip ourselves to move in exactly the opposite direction. In contemporary conditions this requires courage, as much in study as in action, aiming to lay the foundations for the reconstruction of a geographical space that would truly be the space of man, the space of all people, and not space at the service of some, and of capital.

To reach this goal we must understand the fundamental value and the ultimate purpose of the geographical thing. We must pass through the empirical to arrive at the philosophical. Only thus will we realize the ideas of Saint-Simon and Owen (Prestipino 1977, 14). Before us, they were already conscious of the liberatory potential of science in general and of the science of nature in particular.

As nature has become social nature, geography must analyze and define how the conscious use of space can restore dignity to man.

With other social scientists, geographers must lay the foundations of a truly human space. This will be a space that unites men through and for their work, that does not divide classes from one another and exploiters from exploited. It will be a material space worked upon by man but a space that does not turn against him. It will be not an artifice but a social nature that is open to the direct contemplation of human minds. It will not be space as a commodity labored on by another commodity—artificial man. It will be space as an instrument for the reproduction of life.

Acknowledgments

Between 1974 and 1977, I worked at the Universities of Dar-Es-Salaam, in Tanzania, the University of Venezuela, in Caracas, and Columbia University, in New York. My many discussions with students made a huge contribution to the development of the ideas explored here. Many colleagues, similarly, helped me to develop these ideas. I especially want to thank Doctors Antonia Déa Erdens and Maria Auxiliadora da Silva, both professors at the University of Bahia, for their invaluable support and collaboration in the final editorial phase of writing this book.

Notes

Translator's Introduction

1. I have briefly discussed the challenge of translating *técnica* elsewhere, in relation to Melgaço's, Clarke's, and Prouse's work (Davies 2018). It is worth returning to because of its centrality to Santos's theoretical project.

Introduction

1. "only works that reveal to the reader what they have thought for a long time 'without knowing it,' that make them conscious of the implications of their own view of the world, can retain their influence and their agency over the long term" (Goldmann 1968, 40–41).

2. [Santos cites this from Russell (1966) but gives no entry in the bibliography for this date. I have not been able to locate the source of the quotation. —Trans.]

3. Cited in Ribeiro (1968).

4. "Clearly, similar approaches can easily become unpopular, at least in as much as they can seem to come from a place of arrogance. Those who do this can give the impression of wanting to be the 'definitive interpreter of past endeavors and the only guide to the future', and 'to criticize the work of others, in a more or less appreciative way, to at least strive to be constructive'" (Secchi 1968, 17–99).

5. "The composition of this book has been for the author a long struggle of escape, and so must the reading of it be for most readers if the author's assault upon them is to be successful,—a struggle of escape from habitual modes of thought and expression. The ideas which are here expressed so laboriously are extremely simple and should be obvious. The difficulty lies, not in the new ideas, but in escaping from the old ones, which ramify, for those brought up as most of us have been, into every corner of our minds." J. M. Keynes, preface to the English edition of *General Theory of Work, Income and Money* (London: Macmillan, 1942).

1. The Founders

1. Taking into account what Sauer said (1931, 1962, 132) of Camille Vallaux, for whom "the object of geographical investigation is the transformation of natural regions and their substitution for new, already profoundly modified, ones. Camille Vallaux considered the new landscapes created by human labour to be more or less deformations of the natural landscape, and considers the degree of this deformation to be the true measure of the power of human societies."

2. "It was through the appeal of the wider world that the next surge of interest in geography arose in the 1880's—Africa, the New World, a partially known Asia, not to mention the polar seas, all attracted vast interest as nations in the surge of industrialization sought new economic, and sometimes political, conquests beyond their home territory" (Freeman 1961, 48).

3. "An important reason for the lack of recognition accorded to geographical work is therefore the partial nature of the solutions that are offered, in which the dominating influence of the natural environment is implicitly assumed" (Chisholm 1966, 15–16).

4. The accusation against sociologists is made gently by Sorre: "Nevertheless, I fear that among sociologists there pertains unconsciously a tenacious remembrance of geographical determinism. More precisely the physical determinism of E Huntington and that his denunciation has not been accounted for" (1957, 155)

5. "Using the culture concept wherever possible, and welcoming all the help he can get, the cultural geographer surveys a world-wide panorama of man's work asks Who? Where? What? When? And How? The themes of culture, culture area, cultural landscape, culture history, and culture ecology respond to these queries. The geographic study of culture exposes challenging problems, suggests procedures for their solution, and opens the way to an understanding of the processes that have created and are creating new environments for man" (Wagner and Mikesell 1962a, 24).

6. "The major problems of cultural geography will lie in discovering the composition and meaning of the geographic aggregate that we as yet recognize somewhat vaguely as the culture area, in finding out more about what are normal stages of succession in its development, in concerning itself with climactic and decadent phases, and thereby in gaining more precise knowledge of the relation of culture and of the resources that at the disposal of culture" (Sauer 1962, 34).

7. "The geographic complex appears as an assemblage of elements of various ages, each with its own history—and it is not by accident that we use the word 'element,' which carries the same meaning in the vocabulary of plant sociology" (Sorre [1953] 1962, 46).

8. "The ecological approach to human communities is valuable: but too many geographers had assumed that human life is a function of environment and given

too little weight to other factors. In other words, geographical regions bear too plainly the stamp of geographical determinism" (Grigg 1967, 471).

9. "The observational sciences are normally put into three categories: a) The systematic sciences which study things from distinct points of view; botany, for example, is a systematic science; b) Chronological sciences that contemplate a succession of events in time, such as geology; c) Chorological sciences that take space and its subdivisions as a subject, for example geography. In this way, geography's own object is chorological; this is what distinguishes it from neighbouring sciences, whether systematic or chronological. . . . Following the principle of classification is necessarily very different in plant geography and in botany" (Hettner, *Das Wessen und Methoden der Geographie,* cited by Michotte 1921).

10. "Man utilizes the physical milieu through the intermediary of a particular civilization" (Dickinson 1969, 258, quoting Gorou). "The landscapes that the geographer analyses are not ecosystems, but constructions influenced by civilizations and transformed by them . . . the human landscape . . . is explained above all by civilizational factors" (Gourou 1973). This strengthens the cultural relativism of Sauer (1963) and the observations that this tendency elicited from H. C. Brookfield (1964) and Harvey (1969, 11). But there is a distinction to be made between technics understood as the inheritances of a culture, after Gourou (1973) and others and technics as a form of local and partial realization of a mode of production in specific geographical and historical contexts.

11. [Santos appears to have mistaken a page reference here, and I have not been able to identify the reference in published translations of Manuel Castells. To translate, Santos's footnote reads: "For Castells (1971, 57) 'the attempt to explain territorial collectives on the basis of an ecological system has been, up to now, the most serious endeavour to establish a (to a certain extent) autonomous theory according to the optics and logics of functionalism.' On this I recommend G. A. Theodorsen (1961)." —Trans.]

The word "ecology" was coined by Ernst Haeckel (1876, vol. 2, 354), who defined it as the science of the "the correlations between all organisms living together in one and the same locality, their adaptation to their surroundings."

12. The use of the expression "natural resources" is mistaken, but it is difficult for geographers to reclaim from other specialists the use of phrases such as "geographical environment," "physical environment," "natural environment," or even "environment." Among geographers, therefore, ambiguity is the norm. It is curious that even Marxists have not taken care to give words a singular accepted meaning, in spite of successfully defining reciprocal relations through history between a "natural" man and a "socialized" nature. Geography, however, has developed since Marx and—albeit with many exceptions—continues to use the word "nature." I prefer *human space,* or simply *space.*

13. A good critical study of the concept of the region has been made by Darwent Whittlesey (1957).

14. "In the first place, it is the nature of geography—and this makes it irreplaceable—to confer its whole attention to the essential unity of spaces. Geographers are conscious that relations between a group of people and the slice of earth that they occupy in a given environment are, inevitably, effected by other spaces near or far away and at a broader geographical scale than the immediate neighbourhood of the group" (Sautter 1975, 239).

15. "The argument is that the practitioner may find analogies useful in building a model or a theory. This means that he takes concepts from another theory, perhaps from another discipline, and uses them. He will then interpret these concepts in terms of his own theory and, for him, they will only have meaning within such an interpretation and within the normal processes of verification of the theory" (Wilson 1969, 229-30).

16. "Today we are better aware of the risk involved in using correlations to support a theory" (Kerblay 1966, lxvii). "We need not emphasize, however, the danger of too uncritical adoption of physical analogies to studies of social and human phenomena" (Olsson 1967, 13).

2. Philosophical Inheritance

1. On this subject read the introduction to Martonne's *Treatise of Physical Geography,* particularly 22.

2. As early as 1894, Vidal de La Blache wrote in the preface to his *General Atlas:* "It is in *connection* that the geographical explanation of an area lies. Studied separately, the traits which comprise the physiognomy of an area have the status of a fact; they only acquire the status of a scientific notion if we place them in the chain of which they are part and which uniquely gives them their full meaning. We must go further and recognize that no part of the earth bring with it the same explanation. We only discover with any clarity the action of local conditions if we raise our observation above them, and embrace analogies that lead, naturally, to generalizable terrestrial laws" (Sorre 1957, 40-41).

3. According to Mehedinti (1901, 8), Ritter was influenced by Kant, and Humboldt by Comte. The geographical principles of de La Blache and Jean Brunhes are directly inherited from Ritter as much as from Humboldt.

4. "Social thought from about 1870 and 1900 was dominated by Darwin. In England and America Herbert Spencer and in France René Worms helped popularize organic analogies in the social sciences, and these retained vitality in geography long after they had been abandoned in other branches of human studies" (Stoddart 1967, 515).

5. The positivists of Lenin's time rejected the objectivity of space and time. Contrary to Kant, they also did not accept the a priori nature of these notions. As Mach wrote, "without experience in physics, the geometrician would never have reached this conception." "Space and time stand ... for special kinds of sensations" (1914, 345, 348).

6. Sauer was no less an inheritor of positivism, for instance in the importance he gave to "visible" facts, physical or otherwise.

7. On the influence of Kant on geographical thought, see the classic work by May (1970).

8. Still, when we refer to the Kantian conception of space, it's necessary to be clear about which Kant we are talking about. Kant took a preliminary position, considering space as a "phenomenon of relations among substances" (as we read in Jammer 1954, 132). Around 1730, he was already moving toward a Newtonian idea of absolute space (Harvey 1960, 207), before adopting, in 1770, another version, in which space was "a kind of framework for things and events" (Popper 1963, 179). According to Harvey (1969, 207), this would transform into a transcendental vision of space, toward understanding space as a "conceptual fiction". However, it is the ideas of absolute space and of space as a *container* that have most impressed his geographical readers.

9. Einstein's theory of relativity "if we are going to say that the theory supports Kant about space and time, we shall have to say that it refutes him about space-time" (Russell 1958, 133).

10. The idea of a totally interdependent city and region is continually repeated in spatial teaching and research. Among well-known geographers who have adopted this perspective, almost without nuances, are Mark Jefferson (1939), Chabot (1933), Smailes (1953), Alexander (1954), and Jones (1966).

11. [Santos's page reference is incorrect. I have therefore translated his translation. —Trans.]

12. [I have not been able to identify the original source. —Trans.]

3. Postwar Renovation

1. [This title is in English in the original. Santos footnotes the title: "Among other works that mark out the new tendencies in geography are: David Harvey, *Explanation in Geography,* London, Arnold, 1969; Jacqueline Beaujeu-Garnier, *La Geographie, méthodes et perspectives,* Paris, Masson, 1971; Peter Ambrose (ed.), *Analytical Human Geography,* 2nd ed., London, Longman, 1970; R. Chorley & P. Haggett (eds.), *Frontiers in Geographical Teaching,* London, Methuen, 1965, p. 816; B. J. L. Berry & D. Marble (eds.), *Spatial Analysis: A Reader in Statistical Geography,* New York, Prentice Hall, 1968, p. 512; C. Beard, R. Chorley, P. Haggett, & D. Stoddart (eds.), *Progress in Geography, International Review of Current Research,* vol. 1, London, Edward Arnold, 1969; H. French & J. B. Racine, *Quantitative and Qualitative Geography, nécessité d'un dialogue,* Ottawa, 1971. —Trans.]

2. Thomas Kuhn rejects the idea that science can have advanced through a cautious accumulation of facts approaching closer to reality. Kuhn attributes a central importance in the history of science to the appearance of new paradigms, able to define realities by new schemas. Whenever a new problem appears, new problematics must appear. The paradigm is the problematic through which we

can deal systematically with reality. Paradigms succeed one another in confirming the nature of things and the manner in which they are understood (Kuhn 1962).

3. In terms of "New Geography," read Rimbert (1972): "The efforts of the innovators are oriented by four principle objectives: the search for objectivity (which explains, for instance, the preference for factorial analysis to uncover explanatory factors); to speed up processes of compiling and correlating data and analysis (which explains the turn to information technology); simulation of probable evolutions in terms of various, differently weighted hypotheses (hence the importance of probabilities); and an appeal to other disciplines with experience of working with multiple variables. This last objective of interdisciplinarity offers a corrective to the tendency of analysts to be strictly specialized: it is on the frontiers and borders of different scientific domains that the theorists think they can see the greatest number of new paths" (102).

4. A good presentation of the objectives and methods of the so-called New Geography is given by Christofoletti (1976b).

5. [It is likely that Santos is referring to "Geography, Marxism and Underdevelopment," *Antipode* 6, no. 3 (December 1974): 1–9. —Trans.]

4. Quantitative Geography

1. "The use of statistical techniques permits greater precision in these matters . . . if properly used. . . . The practical and methodological problems of geography are such that the use of statistical techniques is likely to exert a strong attraction" (Wrigley 1965, 17).

"Even if linguistic descriptions frequently constitute the first steps towards the development of a theory, they are less precise, less generalizable, and of less predictive value than mathematical models. Therefore, it should come as no surprise that researchers have used such methods to help them understand and predict the diffusion of innovation" (Kariel and Kariel 1972, 46). [This text is not in Santos's bibliography, and I have not been able to identify the source. —Trans.]

2. "The spatial order in the adoption of innovations is very often so striking that it is tempting to try to produce theoretical models which simulate the process and eventually make certain predictions achievable" (Hägerstrand 1967a, 1–32).

3. [Spearman's coefficient denotes the statistical dependence between rankings of two variables. —Trans.]

4. [The citation is in fact to a quotation within Michael McNulty's piece taken from p. 2 of C. A. Moser and Wolf Scott's 1961 work, *British Towns: A Statistical Study of Their Social and Economic Differences*. —Trans.]

5. For Pinto and Sunkel (1966, 83), not all economic problems can be treated in quantitative terms, those that can be analyzed mathematically are not necessarily the most important, and the use of mathematical methods is not the only route toward scientific rigor.

6. For a critique of quantitative geography, see also Dematteis (1970).

7. In "Marxismo e scienze della natura," *Critica Marxista* 1, anno 10, 222 (1972), G. P. recalls "the opposition between diachronic *formation* and mathematics (including, to some extent, cybernetics), as a 'structural' science of a synchronous system" (Russell [1958], referring to Bergson, writes that "true change can only be explained by true duration—and this includes the interpenetration of the past and the present, not a mathematical succession of statistical states"). [I cannot identify the original Russell phrase in the source cited. —Trans.] Ortega and Gasset already wrote in 1936 (1963, 292) that fashionable science is full of problems left unchallenged because they do not fit with the methods currently in use.

8. For Anuchin (1973, 53) "the introduction of a new *method* does not automatically result in the creation of a new *subject* of research."

9. "The current interest in mathematical-statistical analysis of systems of distribution and their reciprocal action in space expands and refines our concepts of reciprocal relations. However, there is a danger of giving excessive importance to these aspects, such that they will end up curtailing the horizons of geography and reducing it to an abstract science of spatial relations" (Broek 1967, 105).

5. Models and Systems

1. "The first conception of a general system was introduced by Ludwig von Bertalanffy, shortly after the Second World War. Later, the insights of other authors became known, among them W. Rose Ashby. The study of the Society for General Systems Research was very significant in the development of a general theory of systems" (Klir 1966, 29). Among other work see Hall and Fagen (1968); Ackoff, Gupta, and Minas (1962); von Bertalanffy (1968); Lee (1970); Emery (1969).

2. In terms of systems analysis in geography, see among others Chisholm (1967); Wilson (1967, 253–69); Chorley (1962); Berry 1964b; Mabogunje (1970). According to Olsson (1967, 13), "The notion of the spatial system must be related, also, with the general theory of systems as proposed by von Bertalanffy (1951, 1962) and Boulding (1956)." The "analysis of systems" and the "general theory of systems" became a fashion in the school of spatial analysis represented by, for example, Chorley (1962); Ajo (1962); Ackermann (1963); and Curry (1964). Often, however, the authors differ in terms of the interpretation of these ideas. In terms of the "theory of systems," also read Christofoletti (1976a, 43–60).

3. "Quantities, scale, the relations between these quantities and properties that determine these relations—these are fundamental markers of the whole system, independently of the scientific discipline or the point of view from which the system is defined" (Klir 1966, 30). "Constructing models is stimulating because, through their ultrageneralizations, those areas where greater definition was needed become clearer. . . . In sum, the role of models in geography is to codify what existed before and to enable new enquiries" (Haggett 1965, 22–23).

4. [The page reference appears wrong. I have not been able to quote from the original source. —Trans.]

5. See principally Stoddart 1967. For Castells (1971, 57), "the attempt to explain territorial collectivities on the basis of the ecological system constitutes the most serious effort up to today to found—up to a certain point—an autonomous theory according to the optic and logic of functionalism". The author suggests reading, in this respect, Theodorsen (1961).

6. "Ecological studies only make sense when integrated into a general analysis of human settlements: these depend on social factors but are constrained by imperfect dominion of the environment. Seen from this angle it is possible to go beyond old possibilist interpretations: the set of relations with the environment and social relations constitute a system of reciprocal connections. However, if this is not apprehended in its totality, then explanations can only be contingent. And it is the social forces that, in general, are the most susceptible to creating order: for a long time they were neglected by a Darwin-inspired geography for which the essential problem was the study of groups' relations with the natural environment" (Claval, 1964, 111).

7. "laws . . . have reciprocal relations. . . . I will examine all these relations: they form a set that we can call the *spirit of the laws*" (Montesquieu 1951, tom. II, 238).

8. "In contemporary language the term 'model' is used in at least three different ways. As a noun, model implies a representation; as an adjective, it implies an Ideal; as a verb, *to model* means to demonstrate. . . . In scientific use, Ackoff; Gupta, and Minas (1962) suggested that we should incorporate part of each of these three meanings; in the construction of models we create an idealized representation of reality in order to demonstrate some of its properties. . . . These models are necessarily made by the complexity of (the nature of) reality. They are a conceptual proof of our understanding and as such offer the professor an apparently rational and simplified framework, for teaching and research they provide a source of working hypotheses to test against reality. Models do not contain the whole truth but a useful and comprehensive part of it (Society for Experimental Biology, 1960)" (Haggett 1965, 19).

9. In his *Novum Organum,* Bacon described scientific theory as containing "rash and premature . . . anticipations of Nature." Certainly, we can agree that many of the models described in the first half of this book fit this description closely: they are all crude, full of exceptions, and easier to refute than to defend. Why, therefore, we should ask, do we work to create models, rather than directly studying the "facts" of human geography? The responses lie in inevitability, economy, and the impulse to create models:

(a) The construction of models can be imagined because there is no fixed dividing line between facts and fictions. In the words of Schilling "this belief in a universe of real things is merely a belief. It is a belief with high probability, certainly, but a belief none the less" (Schilling n.d., 394A).

Models are theories, laws, equations or suspicions which materialize our beliefs about the universe that we think we see.

(b) The construction of models is economical because it allows us to pass on generalizable information in highly condensed forms.

10. In economic geography, the construction of models proceeds along two distinct and complementary paths. In the first, the modeler disaggregates a problem, to begin with simpler postulates, and gradually introduces greater complexity, approximating real life more and more closely. This was Thünen's (1826) insight in his model of the use of the earth in *Der isolierte Staat*. The second method is to begin from reality and make a series of simplifying generalizations. This was Taaffe's insight (Taaffe, Morrill, and Gould 1973, 240–54).

11. "The model only enables complexity because it simplifies. The overall image of complexity reproduces complexity and is thus useless. It is through highlighting a particular character for its importance that we can see the model progress. It is, by its nature, partial and simplifying" (Preliminary reflections on the search for a method of approaching planning studies made by a group of engineers of the *Génie Rural, des Eaux et des Forêts,* France, November 1967).

6. The Geography of Perception and Behavior

1. Jakubowsky (1971, 118) says as much very clearly when he writes that the relation of consciousness to being can be correctly understood when being is understood dynamically, as a process. It thereby gains its rigid form of objectivity: particular things on the surface of social being should not be taken in isolation but conceived as processes within the frame of the social totality.

2. With regard to the proposition of the problem of perception in geography, see de Oliveira 1977. Tucey (1976) offers an exhaustive and high-quality study of the problems of perception and behavior. A critique of the use of theories of perception in geography is made by Riesner (1973). A recent edition (1974, no. 3) of the French journal *L'éspace géographique* is entirely dedicated to the problem of the perception of space. The pieces form an inventory of studies done in various parts of the world connected with a critical overview. Writers include Paul Claval, Vincent Berdoulay, Roger Brunet, Renée Rochefort, Antoine S. Bailly, Jean-Luc Piveteau, Alain Metton, and Armand Fremont.

More recently, the School of Geography of the Universidade de los Andes (Merida, Venezuela) conducted a study of this question, from the point of view of underdeveloped countries and the particularities of the organization of space in the Third World. This constitutes a critical contribution to work on the perception of space undertaken by geographers in the United States and in Europe (Wettstein, Rejas Lopes, and Valbuena 1976).

3. "The subject does not belong to the world; on the contrary, rather, it is a limit of the world" (Wittgenstein 1969). In spite of everything, I am of the opinion that there never has been nor ever will be an objective science of the spirit,

nor an objective doctrine of the psyche. Objectivity consists in condemning psyches to non-existence, submitting them to the forms of space and time" (Husserl 1975, 28). "The individual who perceives an association between actual entities is himself a form of the ultimate creativity of the universe. He is a reflection of the universe of which, as an entity, he could never be independent. He is the universe in this position" (Leslie 1961, 131). "It seems impossible henceforth to proceed with certainty of what is perceived; seen from outside, perception slides over things without touching them. At most it can be said that each of us has a private world, if we desire to impose a perspective of perception on it, these private 'worlds' are worlds only for those who believe in them, but they are not the world. The singular world,...the real world,...is not only that which is glimpsed by our perception" (Merleau-Ponty 1964, 24–26; 1971, 111).

4. See, among others, Pred (first part 1967, second part 1969). For a critical study of behavioral geography see Downs (1970).

5. "In contradistinction to the culturally conscious man of the early humanistic writers, and the efficiency-conscious man of the spatial tradition, man in the behavioural movement is regarded (albeit euphemistically) as a 'decision maker', motivated by diverse and often conflicting personal and psychological drives" (Buttimer 1974, 23).

6. According to Schmidt (1971, 116) for psychology to become a substantial science requires turning to Marx to overcome the separation between psychology and the history of production. For instance, the work of "psychologist S. L. Rubinstein has demonstrated the dependence of the world of perception and the modes of perception on the forms taken by man's activities towards natural objects." For this psychologist, "the specifically human forms of perception are not only a precondition of specific human activity, but also its product."

7. Fraise (1976, 2) alerts us to a recent tendency in psychology that geographers should take note of. "A new tension," he writes, "divides psychologists. For some, the qualification 'behaviourist' has almost become an insult." Those who want to hold on to this tag call themselves neobehaviorists, followers of a subjective behaviorism such as that of Miller, Galenter, and Pribram (1960), of a social behaviorism such as that of Staats (1975), or even of a "mental" behaviorism (Paivio 1975). Many other authors refuse all kinds of behaviorism, seeing it as worth applying only to animals. These call themselves cognitivists or Freudians, existentialists or humanists, but in spite of the vast differences among them they have a common preoccupation: to return to the human.

7. The Triumph of Formalism and Ideology

1. "Recognizing—late—the clear characteristic that this science has often functioned with a positivist, naturalist conception, it occurred to some geographers to endow it with logical instruments of analysis based on mathematical models. The resulting logical positivism achieved some success in the sense that

it believed simultaneously in overcoming the impasse and in usefully intervening in social reality. But the perspective ended up being abandoned by many of
its followers." So wrote Corrêa da Silva (1976a, 93), who mentions David Harvey
as one of many geographers who changed track as a result of a disenchantment
with quantitative geography.

Quantification is also an inheritance of Darwinism, argues Ginsburg (1973, 2).
"Measurement and quantification, those nineteenth century reductionist consequences of social Darwinism, become the fetishistic *sine qua non* of acceptable
scholarly enterprise." Naturally he is speaking about the United States.

2. Modernization, whose exact meaning escaped many geographers in the
West as much as in the Third World, has recently been analyzed in penetrating
detail by Drysdale and Watts (1977). Beyond simply presenting a strong critical
bibliography, their analysis outlines the dialectical process of modernization based
on concrete cases. The authors propose that we take into account both the ideological tools used to demarcate some people as "primitive" under the contemporary world of production and also the different ways in which these peoples
respond.

3. "Natural science methods, particularly quantification, have brought un-
doubted gains. Often, however, the great successes of natural science are cred-
ited to empiricism, which in fact has been responsible for few major advances;
and quantification in the quest of similar successful status for social science is
often premature and misguided- spurious precision with hard numbers repre-
senting soft and isolated facts. When the quantities are cash values in policy-
oriented research, the results can be as ludicrous as the assumption in the Roskill
Report that a man's value to 'the community' is £4,360 while an 'average woman'
is *negatively* valued at £1,120!" (Anderson 1973, 3).

4. "In response to this *prescriptive scientism,* as we called it earlier, geography
has become concerned with, amongst other things, 'law' seeking, model building,
and the articulation of theory: few seem to question whether scientism is appropriate to the study of geography's principal focus—man. In fact no 'laws' have been
produced in geography, and probably never will be. Scientism can make some
descriptions more exact, like Berry's (1964a) approach to regional analysis, but it
is of no explanatory value, no aid to understanding" (Hurst 1973, 43).

5. "As soon as mathematics is applied to a field of problems for which it is
still too weak, there is a risk of creating illusory knowledge, scientific phantoms.
There is also a risk that without knowing or wishing it, i.e. *with no ideological
intent,* the invisible but real line which always separates scientific knowledge from
ideology will have been crossed" (Godelier 1966a, 857). In terms of the relations
between ideology and geography, read Sodré (1976) and the commentary by
Corrêa da Silva (1976a). A theoretical and epistemological analysis of the problem has been undertaken by James Anderson in his classic article "Ideology and
Geography: An Introduction" (1973).

6. As Armstrong (1973) put it, quantitative techniques allow us to measure only what can't be changed.

7. Corrêa da Silva (1976b) suggests that reading some geographical works gives the impression of a mere description of external factors or the *appearance of reality*. [The text in Portuguese that Santos cites as a translation of Slater does not match either of Slater's 1975 texts that he cites in the bibliography. However, it is very close in content to Slater's "The Poverty of Modern Geographical Enquiry" (particularly 160–61). That paper notes that it was originally prepared for a staff-student seminar at the University of Dar es Salaam in August 1974, where Santos and Slater were colleagues. It seems likely, therefore, that Santos is quoting from a version of that paper. I have translated the words Santos gives to Slater from the Portuguese. —Trans.]

8. Hodder and Lee (1974) recognize that geography and regional science had common foundations in neoclassical economics.

8. The Balance of the Crisis

1. "Perhaps the most immediate inertial problem faced by geographers is the constriction imposed by geography's past growth. We inhabit a Victorian academic structure every bit as solid and constraining as its architectural counterpart. Despite the vigorous growth of the subject in universities and schools over the last fifty years, the everyday popular image of geography is an antique one dogged by 'exploration, description and capes and bays' and this in turn influences the character of our intake of young minds, together with the amount, sources and destination of research funds" (Haggett and Chorley 1965, 375).

2. On this question read the important work of Dubarle and Doz (1972).

3. "In the United States evolutionism was more or less identified in biology with Darwin's theories and in sociology with those of Spencer. Not long ago it was enough for a university professor to be suspected of 'Darwinism,' a theory held to be against the teachings of the Bible, for that professor to be stripped of their post" (Cuvillier 1953, 109–110).

4. In a well-documented critique of Friedmann's version of this model, McCall (1962, 8) also critiques the use of abstract empiricism as an approach in spatial studies.

5. If space was always the preferred and indispensable vehicle of capital, after almost a century it has become the object of an ideology: planning. In scientific guise, this ideology developed to facilitate the dominium of capital over space and with the objective of changing society through organizing behavior and the distribution of material things so as to organize society according to the structure of capitalism. But can it be that space, the result of social labor, can be used in no other way than in the service of capital? We will be able to find answers to this question only if we consider space and society as part of a unitary historical process.

6. [In English in the original. —Trans.]

7. "Geography, instead of attempting to be the complete mistress of its own domain, and to embrace as truly its own what rightfully falls within its precincts, surrenders its own resources to other sciences, for them to work and develop" (Ritter 1863, 283. Written 14 April 1836).

8. "it is not surprising, but it is certainly disconcerting, that not one of the recent books on the environmental crisis [sic] has been written by a geographer. It is not surprising, but again it is deeply disconcerting, that on regional development it is the economists and not the geographers who dominate the literature" (Franklin 1973, 207).

9. The citation is found in Fischer et al. (1969, 284).

10. "Specialization ... disguises the profound complexity of human reality, and dampens the curiosity of the researcher" (Dresch 1948, 91).

11. "A conditioning assumption of all bourgeois modes of thought and analysis is the belief in the theoretical and practical impossibility of comprehending and explaining the totality of social reality" (Slater 1975, 168–69). This suggests a fragmentary knowledge that gets further and further away from a global vision while the object of analysis gets ever narrower and more subdivided.

Marchand (1972, 95) wrote: "Uni- and multi-variable methods of analysis are a powerful aid to geographical research, but they are not characteristic of it. They are applied to a series of observations and if, twenty years from now, they become the general rule in modern geographical studies, we must confess that economists and psychologists doubtless employ them better. The fact that geographers work on space provokes, on the contrary, a series of unique methodological problems: the macrogeographic school tries to deal with space as a whole; on the other hand, Dacey's studies and studies on 'filtering' take up an analytical approach; finally, all statistical study puts forward the fundamental problem of 'spatial autocorrelation.'

Contrary to macrogeography, the diverse filtering methods end up dividing geographical space into its elementary components, to better understand the law of the distribution of phenomena."

12. Strabo wrote that "geography must be written to serve Statemen and the dominant classes." He was expanding on an idea from Polybius.

13. Sorre himself was a victim of this. He wrote the most important oeuvre of geographical literature (that I know), but much of his work is practically unknown by young—and less young—geographers. The fashion for manuals and the geography of the mass media, critiqued by Lacoste (1976), is one of the causes of this, but even in France he is underrecognized.

9. A New Interdisciplinarity

1. "Geography brings together all the sciences, opens all horizons, contains all human knowledges" (Faure, "Les enseignements de la guerre," cited by Febvre, *La*

terre et l'évolution humaine, 24). The same author writes: "We place the French university at the top of a pyramid and, right up there with it, the word *geography,* towards which all human knowledges should be directed."

2. "There can be no question that geography as a discipline always has had pretentions to grandeur in synthesis. The trouble is that it has never really developed the tools to bring this about" (Harvey, in Graves and Moore 1972, 41). For more on geographers' superiority complex read Claval 1975, especially 278.

3. [Santos does not give a page reference, and I cannot find the original in the 1973 or 1975 Brookfield references he provides. —Trans.]

4. "In the post-war years, the social sciences in France went through a vigorous readjustment; a reaction set in against the narrow teaching which had prevailed so long, and many young sociologists and economists went to the United States to complete their education. This example was not, however, followed in geography. Right up to 1968, the majority of geographers were of the sincere opinion that, outside France, there was no school of geography worthy of the name" (Claval 1975, 260).

5. "Our search for a professional identity led to an intellectual independence and eventually to a degree of isolation against which a number of the rising younger generation of geographers have now reacted.... In our desire to make our declaration of independence viable, we neglected to maintain a view of the advancing front of science as a whole. We acted as though we did not believe in anything more than the broadest generalities about the universality of scientific method. In effect we neglected to appraise continuously the most profound current of change in our time. We neglected an axiom: The course of science as a whole determines the progress of its parts, in their greater or lesser degrees" (Ackerman 1963, 431–32).

6. "In the French university they try to avoid, wherever possible, reading any language other than that of Descartes" (Goldblum 1974, 138–39).

7. The structure of higher education in France in fact impedes students from being economists and geographers or sociologists and geographers at the same time. The renewal that has been so remarkable in economics and sociology has not had an echo in geography.

8. "If, in the past, there was an interest in forms, phenomena and facts in general, or specific to each scientific field, in our period it seems, on the contrary, that there are universal aspirations. There is a need to discover distant frontiers, and a superimposition and interpenetration of different disciplines according to spatial, physical, organic and intellectual dimensions which later return to a stable equilibrium" (Ritter 1974, 79).

9. This concern with the relations between geography and the social sciences already existed in 1934 for Isaiah Bowman, the American geographer, in a paper that he presented to the American Historical Association. The relations between ethnology and our science have been analyzed by Leroi-Gourhan (1943, 14–19).

This article is a review (P. Deffontaines wrote another), and Leroi-Gourhan attempted to show the intimate relations between the two disciplines, which, he thought, both concerned themselves with questions of space. He wrote: "since human geography is the closest to us, one day a union will need to be made" (19).

10. "Every special science has to assume results from other sciences" (Whitehead 1938, 136).

11. "It must be clear to everyone that the geographer, whether in research or in education, is not seeking to take the place of the economist, the sociologist, the demographer or the ethnologist. But he must know how to use results with full knowledge of cause. That is to say, he must know how they were obtained, and consequently their degree of accuracy" (George 1958).

"It is often noted that geographers demonstrate, in the framework of their fieldwork in this or that region, that certain fashionable economic theories are disproved by facts on the ground. Like M. Jourdan, they undertake—without necessarily knowing that they are—to rectify economics! . . . This is also the case for economists who do fieldwork and refer to geographical phenomena" (Santos 1968).

12. "a biologist, J. Constantin, was already thinking in 1898 about the relations between biology and geography: 'it is in the confines of scientific fields that new problems are located, and which find unexpected and interesting solutions'" (Tulippe 1945, vol. 1, 75).

13. "The mathematical description of the world emerged at an important moment. There are sciences of plants and animals that study living beings and their structure, physiology, behaviour and genesis. These sciences engage with the totality of their activity and their distribution across the earth. The geography of plants or animals is merely one chapter in botany and zoology, worked on by particular specialists, but it is a crucial chapter. What can we say of rocks, their genesis, their distribution and the forms of the earth of which they are the material itself? Can it be that geology, the old 'geognosia' has not long been freed of the ties that could bind it to geography? Can it be that knowledge of the earth—paleontology, mineralogy, tectonics and, when necessary, new, high-precision sciences grouped around the physics of the earth—does not constitute a perfectly realized body of knowledge, and that its progress involves general physical geography? It is necessary to properly emphasize that general physical geography was the condition for the development of this knowledge.

It is no wonder that geographers themselves have been restricted by this situation. They recognized the proliferation triggered by the diversity of geographical material, and were intimidated by the difficulty of accounting for the results of so many disciplines, and the growing number of techniques. This last point is particularly important because novel techniques exercise a powerful attraction on young researchers" (Sorre 1952).

14. For Harvey (1969, 122), Brookfield's great critique of the Berkeley school is of its failure to find a deeper explanation that crosses disciplinary frontiers. This is a difficult subject which can lead geographers to a kind of "intellectual dandyism," offering only superficial interpretations inspired by a poor understanding of related disciplines.

15. "Geography and history fill up the entire circumference of our perceptions: geography that of space, history that of time" (Kant 1802, vol. I, 6). George has recently (in *Geografia Ativa*) written that "geography is the continuation of history" and the geographer is a "historian of the present." Note that Pailhé, Grataloup, and Lévy (1977, 46–47), explicitly recognize their Vidalian affiliation.

16. "While collecting my ideas on the mobility of the *oikoumene* for a study of human migration, I was led to reflect again on the role of historical explanation in human geography. It seemed to me that the terms of the old controversy between historians and geographers were much too restrictive and that it would be rewarding to consider the problem in all its generality. In order to achieve this goal, we must deliberately disregard pedagogic considerations, which have dominated and distorted the debate. We are concerned with questions of method and scientific philosophy; particular academic disciplines are not the issue here" (Sorre [1953] 1962 44).

17. When dealing with space as human space, we should recall what Graves and Moore wrote: "the events of history must happen in some space or place whilst the places of geography exist and evolve through time" (Graves and Moore 1972, 20).

18. "the theoretical construction of traditional social physics is so schematic that it is incapable of interpreting the complexity of real facts" (Claval 1972, 120–21).

19. Tucey (1976) reminds us that the first person to understand the importance of phenomenology in the study of geography was Carl Sauer, who, in his article "The Morphology of Landscape," wrote: "the task of geography is conceived as the establishment of a critical system that embraces the phenomenology of landscape, in order to grasp in all of its meaning and colour the varied terrestrial scene."

20. "The different social and human sciences may be different realms, in whose borderlands trespass is dangerous save for the genius ... a social or human scientist may profit by studying disciplines other than his own. It is dangerous to practise them without training and appropriate skills" (Devons and Gluckman 1964, cited in Harvey 1969, 123).

21. On the consequences of Einstein's work for the progress of the sciences in general, Whitehead (1964, 164) argued that "there is a general agreement on the fundamental merit of Einstein's investigations, irrespective of any criticisms which we may feel inclined to pass on them."

10. An Attempt to Define Space

1. Cited by Boyé ([1970] 1974, 8).

2. Allix (1948): "Geography, we might modestly assert, is the study of the distribution and coordination of facts that exist in the portion of the earth's crust and atmosphere accessible to men." Martonne: "modern geography studies the distribution on the surface of the earth of physical, biological and human phenomena, the causes of their distribution and the local relations of these phenomena. It has an essentially scientific and philosophical character, but is also descriptive and realist." Cited in Tulippe (1945, 2nd part, vol. 1, 80). For Fr. Ruellan (1943), "geography is a science that seeks to define the association between facts in a synthetic manner, in order to better understand their complex relations, that is, to understand a coherent set of manifestations of physical and human life on the surface of the earth. It is worth, therefore, clearly marking out the extension of the phenomena that make up the geographical environment, to seek out their causes and effects, and to trace their evolution." For Cholley (1951), "the question is to know the genesis, structure and evolution of the combinations that can be understood by scientific knowledge. That is, in sum, those that can be measured. For combinations of human geography, this does not raise concerns: a combination is measured by its effects: production, demographic coefficient, quality of life, etc. We can, therefore, fix the moment in which it appears and follow its evolution. For combinations of physical geography, this is equally possible" (77); "We have sought to reduce geography to a way of considering things, to a simple state of mind. We end up seeing that it can represent an order of knowledge. It has its own field, reality and method that belongs to it" (25); "A young science, or a science of the future, as geography seems to us to be" (78).

3. For Hayek, the object of scientific study is never the totality of observable phenomena in a given place and time but always and only some of its abstract aspects. For him, according to Kosik ([1963] 1967, 62), the human mind cannot embrace wholes, or the totality of the diverse aspects of a real situation.

4. [The dates of Santos's references here to Durkheim clash with the texts in his bibliography, and I have not been able to identify the original source for this quotation. —Trans.]

5. "Geography cannot dedicate itself to 'men' or 'the world' in general. It must limit itself to what is specific to it; to the space to be explained and theorized. The field of scientific geography is the point of departure for this" (Lévy 1975, 58).

6. Categories—or, as Mandel (1975, 39) prefers, "fundamental variables"—each acquire a different value depending on the angle by which the *phenomena* (the appearances?) are studied. If providing an explanation or defining an essence is the point of its "analysis of the whole"—which, according to Mandel, no one, apart from a very few, have achieved—then we should note that different

phenomena and their particular aspects give particular variables greater explanatory weight, depending on the historical epoch.

7. [No specific source is cited by Santos for this quotation. —Trans.]

11. Space

1. According to Knox (in Hegel 1962, 313), the best appreciation of Hegel's philosophy of nature in English is the article by S. Alexander in *Mind* from October 1886. An important study of the relations between Hegelian thought and geography was recently written by the French philosopher François Chatelet in *Herodote*, no. 5 (1977).

2. [Santos's page reference here seems to be incorrect, and I have not identified the source in published English translations of Hegel. I have therefore translated Santos's translation and retained the incorrect reference. —Trans.]

3. [Santos's page reference here seems to be incorrect, and I have not identified the source in published English translations of Hegel. I have therefore translated Santos's translation and retained the incorrect reference. —Trans.]

4. Still, it is to Knox (1942, 1962, 305), Hegel's English translator, that we owe an exegesis of Hegel's thought about the Idea. Understanding it would require, on the one side, a set of organically connected ideas and, on the other, a series of natural phenomena and human institutions that includes a set of ideas. The true realization would be the synthesis of the two. Alongside this there are the inseparable accidents and contingencies of the spatial-temporal sphere, with which the philosopher, unlike historians and scientists interested in empirical realities, does not need to concern himself.

5. [In English in the original, a translation of *espaço-continente*. —Trans.]

6. "At the end of this study, we will attempt to respond to the questions we will not judge *a priori*. What is space? What is its place in the process of socioeconomic change? Is it a reflection, a projection of social organization, an element that has no place of its own in the movement of deconstruction-reconstruction? Or does it intervene in this movement" (Vieille 1974, 29)?

7. In terms of the objectivity of spatial phenomena, the Soviet geographer S. V. Kalesnik (1971, 197) wrote the following: "While acknowledging 'things in themselves,' we also hold that the objects of geographical research, namely natural and territorial production complexes, exist objectively, outside of our consciousness, and that man is capable of perceiving these complexes fully through his senses and ideas. This position is entirely in keeping with everyday experience."

8. In a colloquium about the relations between history and geography, the North American historian George Burr acerbically criticized the study presented by the geographer Ellen Semple on "geographical location as a historical factor" and the contribution of another historian, O. G. Libby, on "physiography as a factor of communal life." For Burr, geography—which he confounded with natural facts—could be only one historical factor among others. He insisted on the

fact that inert things do not exercise influence and do not have a causal role. We should not, therefore, attribute to nature what is planned and realized by men.

9. "It is the whole thing that is given, everything that is presented to, or rather, that imposes itself upon, the imagination. It is treating phenomena as things, and treating them as *data* that constitutes the starting point of science," wrote Durkheim in his famous 1900 article *La Sociologie et son domaine scientifique,* reproduced in French by Cuvillier (1953). The original appeared in Italian in the *Rivista italiana di sociologia.* "A fact," wrote Durkheim in his foundational 1895 book *The Rules of Sociological Method,* "can exist even if it is not, or is no longer, useful, whether because it was never directed at a final objective, or whether because, having once been useful, it lost all its utility, and continues to exist merely through inertia." Therefore, he added, "there are more leftovers in society than in biology."

10. "Space is an objective reality," wrote Chesnaux (1976, 157). For Cassirer, "the very *form* of spatial intuition itself bears within it a necessary reference to an objective *existence,* a reality 'in' space" (1953, vol. 1, 203). Jordan (1971, 24) notes that when Durkheim, in his *Rules of Sociological Method* ([1895] 1962, 101–4 and 2) says that social facts are "a category of facts which present very special characteristics: they consist of manners of acting, thinking, and feeling external to the individual, which are invested with a coercive power by virtue of which they exercise control over him," he is referring to the characteristics of society conceived as a totality, as Marx had imagined it. "The pressure . . . is the pressure exerted by the 'totality on the individual.'"

12. Space

1. "In 1950, ¾ of the Brazilian population lived in a coastal slice corresponding to ⅓ of the territory (1.8 million km2). 66% of the population of the country—that is, 47 million people—lived in a coastal strip of 250 km" (da Costa 1969, 17–18).

2. On Brazil, important writings include articles (1964, 1972a) and a book (1972b) by Rattner and an article by da Costa (1969).

3. It is estimated that in advanced industrial countries up to 80 percent of new industrial investments are allocated to the expansion of existing factories.

4. "It is true that in the huge towns civilisation has bequeathed us a heritage which it will take much time and trouble to get rid of. But it must and will be got rid of, however, protracted a process it may be" (Engels 1960).

5. [In English in the original. —Trans.]

6. Wagemann (1933, 13), like Lucien Brocard (*Les conditions générales de l'activité économique*), whom he cites, considers physical elements and territory as conditioning factors of economic phenomena. Marchal says the same (*op. cit.,* tomo I, 31). Engels correctly conceptualizes the geographical place of "second nature" in economic activity. In his letter to Starkenburg on 25 January 1894 he

wrote: "Under economic conditions are further included the geographical basis on which they operate and those remnants of earlier stages of economic development which have actually been transmitted and have survived—often only through tradition or the force of inertia; also of course the external milieu which surrounds this form of society" (Marx and Engels 1964a, 410).

7. [in English in the original. —Trans.]

8. For Rofman (1974a, 18): "The dimension of spatial reality is in a permanent state of adjustment under the influence of economic and social reality." The Center of Studies of Development at the Central University of Venezuela argues that "the social formation of any given country is conditioned at each historical moment by its historical inheritance, by external factors, and by its physical space" (CENDES 1971, vol. 1, 23).

9. [This reference to Lefebvre (1958) is ambiguous. It is not cited in the bibliography, and there are a number of Lefebvre publications of this year. I believe it is a reference to *The Critique of Everyday Life.* —Trans.]

10. "At any one moment the settlement pattern may impose certain kinds of constraints on the succeeding stages of economic development; eg, it determines the scale of the markets for goods and service, the degree to which labour specialization is feasible, and the effectiveness with which capital is employed" (Resources for the Future 1966, 31). "For each new generation the means of labour it inherits from earlier generations become the starting point for further advance, and this is the basis of *continuity in history*" (Kelle and Kovalson 1973, 50–51).

11. "The processes of urbanization involve the creation of a built environment which subsequently functions as a vast man-made resource system—a reservoir of fixed and immobile capital assets to be used in all phases of commodity production and in final consumption. These assets have to be maintained and from time to time renewed if society is to be reproduced in its existing state. A certain proportion of social product has therefore to be laid aside as a surplus to produce the built environment" (Harvey, in Gappert and Rose 1975, 120).

12. "Therefore, we have to ask ourselves, in relation to the historical relationship between space and global society: how was it that the rules of space and the effective occupation of territory responded to the succession and transformation of modes of production which were, throughout history, the effective mechanisms of society. We must also ask ourselves what the role of space was in the social process" (Vieille 1974, 3). "Space, therefore, is always an historical conjuncture and a social form that derives its meaning from the social processes that are expressed through it. Space is capable of producing, on the other hand, specific effects on other domains of the social conjuncture, by virtue of the particular form of articulation of the structural instances that are constituted by it" (Castells 1977, 430). "The environment is really an independent variable, not a constant factor. It is a variable that is transformed by the action of the social and economic systems, but in all cases it is a limiting factor, a set of forces" (Godelier 1974, 32).

13. In Marx and Engels (1969), vol. 1, 42. [Santos appears to be mistakenly attributing this phrase to Feuerbach, when it is Marx and Engel's in *The German Ideology.* —Trans.]

13. Space as Social Order

1. For Harnecker, "in any social formation, with a limited number of exceptions, we find: 1. A *complex economic* structure in which diverse relations of production co-exist. One of these relations occupies a dominant place and imposes its functioning on other relations; 2. A *complex ideological structure,* made up of diverse ideological tendencies. The dominant ideological tendency, that subordinates and deforms other tendencies corresponds, generally, to that of the dominant class; 3. A *complex juridical-political structure,* whose purpose is to fulfil the objectives of the dominant class" (1973, 146–47).

2. On the spatial acting on the social, see Boddy (1976).

3. More recently, Godelier has written against "a 'reductive' economic theory, that is one that, like vulgar materialism, reduces all non-economic structures to the status of *epiphenomena* without great importance in the material infrastructure of societies; and, on the other hand, empirical sociological theories that reduce the whole of society to the consequence of (depending on the case) religion, politics or kinship" (1974, 35).

4. If the economy is "one primordial aspect of men's relationship with nature," writes Garaudy, "in the organic totality of relationships, from which are born technology, science, philosophy, religion and the arts, the economy plays a decisive part; even so it never constitutes the sole driving force, after which everything else is epiphenomenon" (1970, 56).

5. "Created things can be understood under a double register: their internal-internalized unity, immanent to the project in its conception, and their external-immanent unity, the actualization and unity of the completed work, which reveals the knowledge within the ordered whole" (Grisoni and Maggiori 1975, 898).

6. For Durkheim ([1895] 1962, 12), "A legal regulation is an arrangement no less permanent than a type of architecture, and yet the regulation is a "physiological" fact."

7. Harnecker (1973, 115) wrote that "the domination of a particular form of relations of production does not automatically make other relations of production disappear; they can continue to exist, though changed and subordinated to the dominant relations of production."

8. "In philosophy the word 'properties' also has two meanings: the properties of the given thing are manifested, first of all, *in its relation to others.* But the concept of properties is not exhausted by this. Why is it that one thing discloses itself in one way in its relation to others, while another thing will disclose itself differently? Obviously, because this other thing-*in-itself* is not the same as the first one" (Plekhanov 1967, 72). [Santos does not give a page reference to the Hegel

quotation. But the reference in English can be found on p. 246 of the 2010 English translation by George di Giovanni, published by Cambridge University Press. —Trans.]

9. "the opposition between social stasis and social dynamics ... impedes the understanding of what a 'social structure' is, with its various rhythms of structuring, destructuring, restructuring, or total structural inversion" (Gurvitch 1968, 407).

10. [Reference to p. 73 of the English-language edition. —Trans.]

14. In Search of a Paradigm

1. Thus Boulding teaches us (1969, 3) that "we cannot escape the proposition that, as science moves from pure knowledge toward control, that is, toward creating what it knows, what it creates becomes a problem of ethical choice."

2. According to Ritter, "we must ask the Earth itself for its laws" (cited in Hartshorne 1939, 55).

3. [Santos does not give a date or page reference for this quotation, though he includes two texts by Christofoletti in his bibliography. —Trans.]

4. "Although the pervasiveness of the response to this emergent situation certainly tells us that something is happening, its diversity highlights our confusion as to exactly what it is that is happening" (Inkeles 1975, 467).

5. Cited by Woodbridge (1940, 3).

6. For Woodbridge (1940, 15), "understanding is the elimination of surprise. We do not eliminate it; Nature does that." What is it that he wants to understand by this, if not that the understanding comes only from the contemplation of reality and nothing more?

7. "Social structure, theory and technology are interdependent, having evolved in relation to and built on one another. Hence, one cannot be changed without inducing change in the others. A change in technology carries with it changes in social structure and in theory" (Schön 1973, 36).

8. Kuhn (1962) denies that science advances through a careful accumulation of facts that allow an ever closer approximation of realities. Kuhn attributes central importance, in the history of the sciences, to new paradigms appearing with the force to define realities through new schemas. Every time a new problem presents itself, new problematics must appear in parallel. This allows him to deal systematically with the reality he calls "the paradigm." Paradigms succeed one another, such that important changes become verified either in the nature of things or in the way they are understood.

9. "One's conception of the world is a response to certain specific problems posed by reality, which are quite specific and 'original' in their immediate relevance. How is it possible to consider the present, and quite specific present, with a mode of thought elaborated for a past which is often remote and superseded?

When someone does this, it means he is a walking anachronism, a fossil, and not living in the modern world, or at the least that he is strangely composite" (Gramsci 1999, 326).

15. Total Space in Our Time

1. [In English in the original. —Trans.]

2. It is more than a century since Demangeon (*Problèmes de géographie humaine*) spoke of the solidarity that connects nations and tends toward making the world into a single great market. This is an interpretation of the contemporary conditions of geographical space with international economic conditions as a backdrop, as also described by the Venezuelan geographer Ramón A. Tovar (1974, esp. 7–23).

3. "But it should give us pause—perhaps it is even cause for alarm—that so little attention is being paid to the problem of social change at the global level, at least when one expresses the amount of that attention as a proportion of the time the world's specialists in sociology and political economy spend on other scientific problems and issues" (Inkeles 1975, 467).

4. Ecumene: is the word Strabo's or Sorre's? Chaunu (1974) takes up the notion of the *ecumene* while changing the word. He speaks of *full-spaces,* whose definition is interpreted in this way by Sautter (1975, 234): bodies of population at a particular stage of technological development that occupy portions of the terrestrial earth as driving forces of history.

5. "Feuerbach did not see how the sensuous world around him is, not a thing given direct from all eternity, remaining ever the same, but the product of industry and of the state of society; and, indeed, in the sense that it is an historical product, the result of the activity of a whole succession of generations, each standing on the shoulders of the preceding one, developing its industry and its intercourse, modifying its social system according to the changed needs" (Marx and Engels 1947, 35).

6. For Bowman "the natural environment is always a different thing to different groups. Its potentialities are absolute but their realization is a relative matter, relative to what the particular man wants and what he can get with the instruments of power and the ideas at his command and the standard of living he demands or strives to attain" (1934, 115).

16. State and Space

1. On this read Hla Myint (1965), above all 72.

2. Britto (1973) conducts a very interesting study of the imbrications between politics and space, seen from the regional angle. After a good critical overview of the work done by geographers, sociologists, and other social scientists (including, naturally, political scientists), the author turns to a detailed analysis of problems connected to the exercise of politics and its territorial conditions, such as the

effects of political action on the transformation of regional spaces. The concerns of geopolitics itself are also covered. Claval (1968) wrote an important study of the relations between political structures of different levels and the region.

3. See Kayser (1966). Among those who laid the foundations of the discipline are Mackinder (1943); Haushoffer (1944); Gottman (1952). Within the classics, also read the works of Ratzel and Reclus.

17. The Ideas of Totality and Social Formation and the Renovation of Geography

1. This citation is found in Sorre (1957, 50). Sorre thought that geography, as "a synthetic discipline, was constantly at risk of dismemberment."

2. I used the term "spatiology" (1974c). Afterward, I discovered that I was in good company: for Lefebvre (1991, last chapter), spatiology, or space-analysis (247), is the science of the future if we want to understand how earthly habitation will no longer be, as it is now, "a prison for man and his utopias."

3. The expression "spatial formation" was apparently used by Mathieu (1974, 89) to identify homogenous regions according to the forms of city-country relations and the corresponding organization of space.

4. A total society is a *collective macrounity*, completely autonomous and solidly structured and organized. It is an overarching macrogroup that manifests itself internally by its de facto dominance and de jure sovereignty over the agents and groups that it encompasses and externally by its de facto separation and independence from other total societies that surround it (Bancal 1974, 226).

5. On social formations and the application of this category to geographical studies, read Santos (1977c) (French); (1977b) (Portuguese); (1977a, 1978a) (Spanish); and 1977d (English). The edition of the journal *Antipode* that includes the English-language version, edited by Richard Peet and me, is dedicated to the relationship between social formations and the organization of space ("Socio-Economic Formation and Spatial Organization").

6. The idea of the social formation comes from Marx. It was honed by Lenin, when he sought a framework of analysis to study Russian reality at the beginning of the century. But the idea, though very rich, has not progressed far since. The Stalinist period, the centralism of communist parties, and the Cold War hindered its development. The reinitiation of studies of this type, in both theoretical and empirical terms, is a new phenomenon. In Italy, France, Latin America, and, more recently, in England, the United States, and indeed Africa, debate has begun and considerable progress has been made.

7. Lenin insisted on recognizing the global, historical impact of the social and economic formation. At the end of the nineteenth century Antonio Labriola sought to redefine this category of historical materialism, to correct those who thought of it as "an economic and historical interpretation" when it was "a historical conception of the economy." For Labriola (1902, 81) it is "the totality of

the unity of social life" that comes to comes to mind with the idea of the social formation.

18. The Idea of Time in Geographical Studies

1. If it is true that geographers have concerned themselves for decades with the problem of differentiated modernization (see, for example, Brown and Moore 1969), we owe the Swedish geographer Torsten Hägerstrand the systematization of this idea and the construction of a true theory of the diffusion of innovations. His indispensable body of work is based in New Geography (Hägerstrand 1967a). See also Hägerstrand (1965). On this theme also see Brown n.d.; Brown 1958. See too Hägerstrand 1966; Gould 1964; Wolpert (1966); Gould 1969; Hudson 1969; Morrill 1968. I have analyzed the problem of the relations between space and time in an article published in 1972 in *Tiers Monde* and also in Santos 1976. For a more detailed critique see the relevant chapter in Santos 1978c.

2. "The reciprocal relations between events are at once spatial and temporal. If we imagine them to be only spatial we will be overwriting the temporal element, and if we imagine them to be exclusively temporal we will be eliminating the spatial element. Thus, when we think only of space or only of time, we are dealing with abstractions. That is, we will be leaving aside an essential element of nature as we experience it in our sensory being" (Whitehead 1969, 168). [It appears that Santos's page reference is incorrect here, and I have not been able to identify the passage he translates from Whitehead. The quotation is a translation of Santos's footnoted translation. There are various similar passages in Whitehead, for example when he notes: "This misapprehension is promoted by the neglect of the principle that, so far as physical relations are concerned, contemporary events happen in *causal* interdependence of each other" (1929, 84). —Trans.]

3. "I believe that being a geographer basically means to appreciate that when events are seen located together in a block of space-time they inevitably expose relations which cannot be traced anymore, once we have bunched them into classes and drawn them out of their place in the block. (This viewpoint is our way of overcoming the scientific fragmentation of knowledge and it is because of that we never fit well in the traditional classifications of sciences or compartments of academic systems)" (Hägerstrand 1974).

4. "Space and time," says Kant, "are not concepts; they are forms of 'intuition'" (cited in Russell 1945, 708).

5. For Bergson (1968, 169), time and space "do not begin to interlace except at the moment they both become fictitious."

6. Alexander suggests that time and space be examined according to a method that allows us to identify empirical aspects or variables of the world and also enables us to make abstractions and intellectual constructions so that we can

guide its constant and dynamic aspects. This method is called *intuition* (drawn from A. C. Benjamin, 1966, 26).

7. See Santos 1971b.

8. [In English in the original. —Trans.]

9. See Santos 1971b.

10. Broek 1967, 105.

11. Boulding 1956, 66–75.

Conclusion

1. "the conscious perception of a change in a historical period is an active and collective evolutionary factor" (Chesneaux 1976, 133).

2. "New social ideas and doctrines precede social revolution, and are one of its prerequisites. The emergence of new social theories and their effect on the popular consciousness, the realization of science in the sphere of material production and the reform of social relations are merely different aspects of a single process, different pathways to accelerate social progress" (Trapezinov 1972, 65).

3. "the intelligence and initiative of people is not in question. But discoveries cannot be used except when changes in social structure and modes of production threaten the system. There are, on the contrary, periods of stability, in which innovations, sometimes found in the archives, are not used" (Haudricourt 1964, 35).

4. Hegel wrote that the *basis* of interpretation of a phenomenon is its essence. Causes, on the other hand, are the transitory, the ephemeral, the instantaneous. Causes are not permanent, but something that slips past, only appearing in a transitory way in the unfolding of phenomena (Haveman 1971, 137).

5. Althusser (see Harnecker 1973, 151) laments that the theory of the "economic level" is not a complete theory, in part because he is not concerned with its other instantiations. He needs to expand his scope to see that space as an *analytical instance* has been forgotten by Marxism (a critique laid out by Rofman [1973, 1974a] and Lacoste [1976]). It is true, however, that in the past both Plekhanov and Bukhárin, like Ratzel and Reclus, did concern themselves with this same question.

6. In our field this would be called "spatialism," against which Coraggio (1974, 1977, 86), Browne (1972, 73), and Barrios (1976, 24) rightly caution us. Browne refers more concretely to the urban to lament that there are still "social phenomena treated as if they were spatial phenomena." Coraggio (1977, 18) extends his account to space in general and advises "conceptually separating, with the greatest possible clearness, what constitutes material spatial manifestation, from the social structures proper which regulate the production of such phenomena." Following the same train of thought, Barrios (1976, 24) writes against attitudes that lead to approaching spatial structures "as the material product of social processes. Modified space is part of the total rationality of what is called structure, but only in as much as it interferes with the action of men."

7. "Man can act upon the world only by breaking it into pieces—by dissecting it into separate spheres of action and objects of action" (Cassirer 1953, vol. 3, 36).

8. [Santos does not provide an entry for Ferkiss in his bibliography, and I have not been able to identify the source of this quotation, so I have translated Santos's translation. —Trans.]

Bibliography

Translator's note: Titles preceded by an asterisk appear in the original bibliography but are not referenced in the text.

Abdel-Malek, Anouar. 1977. "Geopolitics and National Movements: An Essay on the Dialectics of Imperialism." *Antipode* 9, no. 1 (February): 28–36.

Ackerman, Edward A. 1963. "Where Is a Research Frontier?" *Annals of the Association of American Geographers* 53, no. 4: 429–40.

Ackoff, R. L., S. K. Gupta, and J. S. Minas. 1962. *Scientific Method: Optimizing Applied Research Decisions.* New York: Wiley.

Aguilar, Alonso. 1973. *Economía política y lucha social.* Mexico: Nuestro Tiempo.

Aguilar, Alonso. 1974. "Ha avanzado el Marxismo en los últimos años?" *Problemas del Desarollo* (México) 18: 93–96.

Ajo, Reino. 1962. "An Approach to Demographical System Analysis." *Economic Geography* 38, no. 4: 359–71.

Alexander, John W. 1954. "The Basic–Nonbasic Concept of Urban Economic Functions." *Economic Geography* 30, no. 3: 246–61.

Allix, André. 1948. "L'esprit et les méthodes de la géographie." *Les Études Rhodaniennes* [later *Revue de Géographie de Lyon*] 23, no. 4: 295–310.

Althusser, Louis. 1965. *Lire Le Capital.* Paris: Maspero.

Ambrose, Peter J. 1969. *Analytical Human Geography: A Collection and Interpretation of Some Recent Work.* Concepts in Geography 2. London: Longmans.

Amedeo, Douglas, and R. D. Golledge. 1975. *An Introduction to Scientific Reasoning in Geography.* New York: Wiley.

Anderson, James. 1973. "Ideology in Geography: An Introduction." *Antipode* 5, no. 3: 1–6.

Anderson, Nels. 1964. *Urbanism and Urbanization.* Leiden: Brill.

Anuchin, V. A. 1973. "Theory of Geography." In *Directions in Geography.* Ed. Richard J. Chorley, 43–63. London: Methuen.

Armstrong, Warwick. 1973. "Crítica de la teoría de polos de desarrollo." *Revista EURE–Revista de Estudios Urbano Regionales* 3, no. 7.

Arrow, Kenneth J., Hollis B. Chenery, Bagicha S. Minhas, and Robert M. Solow. 1961. "Capital–Labor Substitution and Economic Efficiency." *Review of Economics and Statistics* 53: 225–50.

Avineri, Shlomo. 1970. *The Social and Political Thought of Karl Marx.* Cambridge: Cambridge University Press.

Bachelard, Gaston. 1947. *La formation de l'esprit scientifique: contribution à une psychanalyse de la connaissance objective.* Paris: J. Vrin.

Bagú, Sérgio. 1973. *Tiempo, realidad social y conocimiento: propuesta de interpretacion.* 2nd ed. Mexico: Siglo XXI.

Baltra Cortes, A. 1966. *Problemas del subdesarrollo económico latinoamericano.* Buenos Aires: Ed. Universitaria.

Bancal, Jean. 1974. *L'économie des sociologues: objet et projet de la sociologie économique.* Paris: Presses Universitaires de France.

Barnes, Harry. 1925. *The New History and the Social Studies.* New York: Century.

Barrios, Sonia. 1976. "Prediagnóstico espacial: el marco teórico." Caracas: Centro de Estudios del Desarrollo (CENDES), Universidad Central de Venezuela.

Barrios, Sonia. 1977. "Political Practice and Space." *Antipode* 9, no. 1: 36–39.

*Barrows, Harlan. 1923. "Geography as Human Ecology." *Annals of the Association of American Geographers* 13, no. 1: 1–14.

Bauer, Peter, and Basil Yamey. 1957. *The Economics of Underdeveloped Countries.* Chicago: University of Chicago Press.

Beard, C., R. Chorley, P. Haggett, and D. Stoddart, eds. 1969. *Progress in Geography. International Review of Current Research.* Vol. 1. London: Edward Arnold.

Beaujeu-Garnier, Jacqueline. 1971. *La géographie: methodes et perspectives.* Paris: Masson.

Benjamin, A. C. 1966. "Ideas of Time in the History of Philosophy." In *The Voices of Time.* Ed. J. T. Fraser. New York: George Braziller.

Bergsman, Joel, Peter Greenston, and Robert Healy. 1972. "The Agglomeration Process in Urban Growth." *Urban Studies* 9, no. 3: 263–88.

Bergson, Henri. 1968. *Durée et simultanéité.* Paris: Presses Universitaires de France.

Berque, Jacques. 1970. *Sociologie des mutations.* Paris: Anthropos.

Berry, Brian. 1964a. "Approaches to Regional Analysis: A Synthesis." *Annals of the Association of American Geographers* 54: 2–11.

Berry, Brian. 1964b. "City as Systems within Systems of Cities." In *Regional Development and Planning, a Reader.* Ed. John Friedmann and William Alonso. Boston: MIT Press.

Berry, Brian, and Allan Pred. 1965. *Central Place Studies: A Bibliography of Theory and Applications.* Bibliography Series (Regional Science Research Institute) no. 1. Philadelphia: Regional Science Research Institute.

Berry, Brian, J. L., Frank E. Horton, and Josephine O. Abiodun. 1970. *Geographic perspectives on urban systems: with integrated readings*. Englewood Cliffs, N.J.: Prentice Hall.

Berry, Brian J. L., and Duane F. Marble. 1968. *Spatial Analysis: A Reader in Statistical Geography*. New York: Prentice Hall.

Bettelheim, Charles. 1970. "Sur la persistance des rapports marchands dans les pays socialistes." *Les Temps Modernes* 25, no. 284: 1417–45.

*Bitsakis, E.-I. 1974. "Relativisme et relativité." *La Pensée* 174 (April): 20–31.

Blaut, J. M. 1971. "Space, Structure and Maps." *Tijdschrift voor Economische en Sociale Geografie* 62, no. 1: 18–21.

Bloch, Marc. 1974. *Apologie pour l'histoire, ou, métier d'historien*. Paris: A. Colin.

Boddy, Martin. 1976. "Urban Political Economy: Introduction." *Antipode* 8, no. 1: 1–2.

Bohm, David. 1965. *The Special Theory of Relativity*. New York: W. A. Benjamin.

Boulding, K. 1956. "Toward a General Theory of Growth." *General Systems Yearbook* 1: 66–75.

Boulding, K. 1966. *The Impact of the Social Sciences*. New Brunswick, N.J.: Rutgers University Press.

Boulding, K. 1969. "Economics as a Moral Science." *American Economic Review* 59, no. 1: 1–12.

Bowman, I. 1934. *Geography in Relation to the Social Sciences*. New York: Scribner's.

Boyé, Marc. 1974. "É a geografia uma ciencia?" *Boletim Geográfico* 33, no. 242: 5–17.

Briceño, Roberto, et al. 1974. *Hacia una teoría materialista del espacio*. Caracas: Escuela de Sociología, Universidad Central de Venezuela.

Britto, Luiz Navarro de. 1973. *Politique et espace régional*. Paris: Ophrys.

*Brocard, Lucien. 1929. *Principes d'économie nationale et internationale: par Lucien Brocard*. 3 vols. Paris: Librairie du Recueil Sirey.

Broek, Jan Otto Marius. 1967. *Geografía: su ambito y su trascendencia*. México: Unión Tipográfica Editorial Hispano Americana.

Brookfield, H. C. 1964. "Questions on the Human Frontiers of Geography." *Economic Geography* 40, no. 4: 283–303.

*Brookfield, H. C. 1973. "One Geography and a Third World." *Transactions of the Institute of British Geographers* 58 (March).

*Brookfield, H. C. 1975. *Interdependent Development*. London: Methuen.

Brown, L. A., and E. G. Moore. 1969. "Diffusion Research in Geography: A Perspective." *Progress in Geography* 1: 119–57.

*Brown, Lawrence A. n.d. "Models for Spatial Diffusion Research: A Review." Evanston, Ill.: Spatial Diffusion Studies, Report 3.

*Brown, Lawrence A. 1968. *Diffusion Dynamics: A Review and Revision of the Quantitative Theory of the Spatial Diffusion of Innovation*. Lund Studies in Geography. Ser. B, Human Geography no. 29. Lund: CWK Gleerup.

*Browne, Enrique. 1972. "La eficiencia de la ineficiencia." *Revista EURE—Revista de Estudios Urbano Regionales* 3, no. 9: 63–88.

Brunhes, Jean. 1910. *La géographie humaine: essai de classification positive, principes et exemples.* Paris: Alcan.

Brunhes, Jean. 1947. *La géographie humaine.* Paris: Presses Universitaires de France.

Brunhes, Jean. 1952. *Human Geography: An Attempt at a Positive Classification.* Chicago: Rand McNally.

Brunhes, Jean. 1956. *La géographie humaine.* Paris: Presses Universitaires de France.

Bryan, P. W. 1933. *Man's Adaptation of Nature.* London: University of London Press.

Bukharin, Nikolaï. [1926] 2011. *Historical Materialism: A System of Sociology.* New York: Routledge.

Bunge, William. 1966. *Theoretical Geography.* Lund: Royal University of Lund, Department of Geography.

Burton, Ian. 1963. "The Quantitative Revolution and Theoretical Geography." *Canadian Geographer/Le Géographe Canadien* 7, no. 4: 151–62.

*Buttimer, Anne. 1971. *Society and Milieu in the French Geographic Tradition.* Chicago: Published for the Association of American Geographers by Rand McNally.

*Buttimer, Anne. 1974. *Values in Geography.* Resource Paper, Association of American Geographers. Commission on College Geography no. 24. Washington, D.C.: Association of American Geographers.

Caillois, Roger. 1964. *Instincts et société: essais de sociologie contemporaine.* Genève: Éditions Gonthier.

Calabi, Donatela, and Francesco Indovina. 1973. "Sull'uso capitalistico del territorio." *Archivo di Studi Urbani e Regionali, Franco Angeli Editore* 2.

Cassirer, Ernst. 1953. *The Philosophy of Symbolic Forms.* Vols. 1–3. New Haven: Yale University Press.

Castells, Manuel. 1971. *Problemas de investigación en sociología urbana.* México: Siglo XXI.

Castells, Manuel. 1976. "Theory and Ideology in Urban Sociology." In *Urban Sociology: Critical Essays.* Ed. C. G. Pickvance. London: Tavistock.

Castells, Manuel. 1977. *The Urban Question: A Marxist Approach.* London: Hodder Education.

Ceceña Cervantes, José Luis. 1970. *Superexplotacion, dependencia y desarrollo.* Mexico: Editorial Nuestro Tiempo.

Centro de Estudios del Desarrollo (CENDES). 1971. *Desarrollo urbano y desarrollo nacional.* Caracas: Universidad Central de Venezuela.

Chabot, G. 1933. "Les courbes isochrones dans l'étude de la géographie urbaine." Congrès International de Géographie, Amsterdam.

Chaunu, Pierre. 1974. *Histoire, science sociale: la durée, l'espace et l'homme à l'époque moderne.* Paris: Société d'Édition d'Enseignement Supérieur.

Chesneaux, Jean. 1976. *Du passé faisons table rase?: à propos de l'histoire et des historiens.* Paris: Maspero.

Chisholm, Michael. 1966. *Geography and Economics.* Bell's Advanced Economic Geographies. London: London Bell.

Chisholm, Michael. 1967. "General Systems Theory and Geography." *Transactions of the Institute of British Geographers,* no. 42: 45–52.

Chisholm, Michael. 1975. *Human Geography: Evolution or Revolution?* Harmondsworth: Penguin.

Chivino, L. S. 1973. "Enfoques de la planificación en Zambia." In *Planificación Regional y Desarollo Nacional en África.* Ed. E. L. Mabogunje, 223–43. Buenos Aires: SIAP.

Cholley, André. 1951. *La géographie.* Paris: Presses Universitaires de France.

Chorley, Richard J. 1962. "Geomorphology and General Systems Theory." *Transactions of the Institute of British Geographers* 42: 45–52.

Christofoletti, Antonio. 1976a. "A teoria dos sistemas." *Boletim de Geografia Teorética* 1, no. 2: 43–60.

Christofoletti, Antonio. 1976b. "As características da nova geografia." *Boletim de Geografia Teorética* 1, no. 1 (April): 3–33.

Claval, Paul. 1964. *Essai sur l'évolution de la géographie humaine.* Paris: Les Belles Lettres.

Claval, Paul. 1968. *Régions, nations, grands espaces.* Paris: M.-Th. Genin/Librairies Techniques.

Claval, Paul. 1970. "L'espace en géographie humaine." *Canadian Geographer/Le Géographe canadien* 14, no. 2: 110–12.

Claval, Paul. 1972. "La réflexion théorique en géographie et les méthodes d'analyse." *L'Éspace Géographique* 1: 7–22.

Claval, Paul. 1974a. *Éléments de géographie humaine.* Paris: M.-Th. Genin/ Librairies Techniques.

*Claval, Paul. 1974b. "Contemporary Human Geography in France." *Progress in Geography* 7. London: Edward Arnold.

Cohen, Ernesto. 1973. *Estructura social y dependencia externa: interacción y variables críticas.* Lima. Septembro.

Coraggio, José Luis. 1974. "Consideraciones teórico-metodológicas sobre las formas sociales de organización del espacio y sus tendencias en América Latina." *Revista Interamericana de Planificación* 8, no. 32: 79–101.

Coraggio, José Luis. 1977. "Social Forms of Space Organization and Their Trends in Latin America." *Antipode* 9, no. 1 (February): 14–18.

Córdova, Armando. 1974. "Fundamentación histórica de los conceptos de heterogeneidad estructural." *Economia y Ciências Sociales* 13, no. 14.

Cornu, Auguste. 1945. "Marxisme et ideologie (i)." *La Pensée,* série 2 (January): 89–100.

Costa, Rubens Vaz da. 1969. "O desenvolvimento regional no Brasil e no mundo." *Revista Econômica do Nordeste* 1, no. 2: 5–19.

Costa, Rubens Vaz da. 1972. *Economic Development and Urban Growth in Brazil.* Rio de Janeiro: Banco Nacional de Habitação/Information Office.

Coutsinas, George. and Catherine Paix. 1977. "External Trade and Spatial Organization: A Typology." *Antipode* 9, no. 1: 97–109.

Croce, Benedetto. 1968. *Théorie et histoire de l'historiographie.* Genève: Droz.

Crosta, P. L. 1973. "I processi di urbanizazione: problemi dell'analisi in funzione dell'intervento sul território." *Note sulla impostazzione e gli argomenti del corso.* Venezia: I. U. A. V. Maggio. (Corso di introduzione all'urbanistica).

*Curry, Leslie. 1964. "The Random Spatial Economy, an Explanation in Settlement Theory." *Annals of the Association of American Geographers* 54: 138–46.

Cuvillier, Armand. 1953. *Où va la sociologie française?* Paris: La Pleiade.

Dalton, George. 1971. *Economic Anthropology and Development: Essays on Tribal and Peasant Economies.* New York/London: Basic Books.

Darby, H. C. 1953. "On the Relations of Geography and History." *Transactions and Papers (Institute of British Geographers),* no. 19: 1–11.

Dasgupta, S. 1964. "Underemployment and Dualism. A Note." *Economic Development and Cultural Change* 12, no. 2: 174–85.

Demangeon, Albert. 1929. "Les aspects actuels de l'économie internationale (deuxième article)." *Annales de Géographie* 38, no. 212: 97–112.

Demangeon, Albert. 1943. *Problèmes de géographie humaine.* Paris: A Colin.

Demangeon, Albert. 1947. *Traité de géographie humaine.* Paris: A Colin.

Dematteis, Giuseppe. 1970. *Rivoluzione quantitativa e nuova geografia.* Torino: Laboratorio di Geografia Economica de la Università di Turin.

Desmond, Gerald. 1971. "Urbanization and National Development." In *Urbanization and National Development.* Ed. Leo Jakobson and Ved Prakash, 57–79. Beverly Hills, Calif.: Sage.

Devons, Ely, and Max Gluckman. 1964. *Closed Systems and Open Minds: The Limits of Naïvety in Social Anthropology.* Chicago: Aldine.

Dickinson, Robert Earl. 1969. *The Makers of Modern Geography.* London: Routledge and Kegan Paul.

*Doherty, Joe. 1974a. "The Role of Urban Places in Socialist Transformation (Some Tentative and Introductory Notes)." University of Dar-es-Salaam, Department of Geography.

Doherty, Joe. 1974b. "Introduction: Geographic Research and Methodology." *Journal of the Geographical Association of Tanzania* 10 (April): 1–3.

Dorfles, Gillo. 1972. *Símbolo, comunicación y consumo.* Barcelona: Editorial Lumen.

*Downs, Roger. 1970. "Geographic Space Perception, Past Approaches and Future Prospects." *Progress in Geography* 2. London: Edward Arnold.

Dresch, Jean. 1948. "Réflexions sur la géographie." *La Pensée* 18: 87.

Drysdale, Alasdair, and Michael Watts. 1977. "Modernization and Social Protest Movements." *Antipode* 9, no. 1: 40–55.

Dubarle, Dominique and André Doz. 1972. *Logique et dialectique*. Sciences humaines et sociales. Paris: Larousse.

Durkheim, Émile. [1898] 1962. *L'année sociologique*. Bibliothèque de philosophie contemporaine. Paris: Félix Alcan.

Durkheim, Émile. [1895] 1962. *The Rules of Sociological Method*. 8th ed. New York: Free Press of Glencoe.

Einstein, Albert. 1954. *Ideas and Opinions*. New York: Crown.

Emery, F. E. 1969. *Systems Thinking: Selected Readings*. Penguin Modern Management Readings. Harmondsworth: Penguin.

Engels, Friedrich. 1960. *Anti-Dühring*. Paris: Sociales.

Engels, Friedrich. 1963. "Lettre à Starkenburg, 25 Janvier 1894." In K. Marx and F. Engels, *Lettres sur Le Capital*. Paris: Sociales.

Estall, R. C. 1972. "Some Observations on the Internal Mobility of Investment Capital." *Area* 4, no. 3: 193–98.

Fackenheim, Emil L. 1961. *Metaphysics and Historicity*. Milwaukee: Marquette University Press.

Faure, Albert. 1970. "Les enseignements de la guerre." In Lucien Febvre, *La terre et l'évolution humaine*. Paris: Albin Michel.

Febvre, Lucien. 1932. *A Geographical Introduction to History*. History of Civilization. London: Kegan Paul.

Ferkiss, Victor. 1974. *The Future of Technological Civilization*. New York: George Braziller.

Ferrari, Giorgio. 1974. "Territorio sviluppo: un comprensorio nella regione Veneta." *Critica Marxista* 12: 79–93.

*Feyerabend, Paul. 1968. "How to Be a Good Empiricist: A Plea for Tolerance in Epistemological Matters." In *The Philosophy of Science*. Ed. P. H. Nidditch, 12–39. London: Oxford University Press.

Fischer, Eric. 1969. *A Question of Place: The Development of Geographic Thought*. 2nd ed. Arlington, Va.: Beatty.

Fraise, Paul. 1976. "Psychologie: science de l'homme ou science du comportement." Address to the International Conference of Psychologists, Paris.

Franklin, S. H. 1973. "Ha Rollo!" *Pacific Viewpoint* 14, no. 2: 207–10.

Freeman, T. W. 1961. *A Hundred Years of Geography*. London: Gerald Duckworth.

French, Hugh, and Jean Racine. 1971. *Quantitative and Qualitative Geography: La Nécessité d'un Dialogue*. Occasional Paper (Department of Geography) 1. Ottawa: University of Ottawa Press.

Funes, Julio Cesar. 1972. *La ciudad y la región para el desarrollo*. Caracas: Comisión de Administración Publica de Venezuela.

Gappert, Gary, and Harold Rose. 1975. *The Social Economy of Cities*. Beverly Hills, Calif.: Sage.

Garaudy, Roger. 1970. *Marxism in the Twentieth Century*. London: Collins.

George, Pierre. 1958. "III. La place de la géographie humaine parmi les sciences humaines. Problèmes de méthode et d'opportunité." *Bulletin de l'Association de Géographes Français* 35, no. 275: 58–63.

Geras, Norman. 1971. "Essence et apparence: aspects du fétichisme dans le *Capital* de Marx." *Temps Modernes* 27, no. 304: 626–50.

Gerratana, Valentino. 1973. "Formación social y sociedade de transición." *Teoria del Processo de Transición (Cuadernos de Passado y Presente)*. *Critica Marxista* (Cordoba, Argentina), 45–79.

Ginsburg, Norton. 1973. "From Colonialism to National Development: Geographical Perspectives on Patterns." *Annals of the Association of American Geographers* 63, no. 1: 1–21.

Goblot, J. J. 1967. "Pour une approache théorique des faits de civilisation." *La Pensée* 133 (June, August, December): 134–36.

Godelier, Maurice. 1966a. "Système, structure et contradiction dans le *Capital*." *Temps Modernes* 246 (November): 828–64.

Godelier, Maurice. 1966b. "Un retour aux problèmes fondamentaux et indispensables, conaissons-nous le fonctionnement des structures sociales?" *Developpement et Civilisation* 28 (December).

Godelier, Maurice. 1969. *Racionalided a irracionalidade na economia*. Rio de Janeiro: Tempo Brasileiro.

Godelier, Maurice. 1972. *Rationality and Irrationality in Economics*. London: New Left Books.

Godelier, Maurice. 1974. "Considerations theoriques et critiques sur le problème des rapports entre l'homme et son environnement." *Social Science Information* 13, no. 6: 31–60.

Goldblun, William. 1974. "Comentário." *La Pensée* 174 (April): 138–39.

Goldmann, Lucien. 1968. *Marxismo, dialectica y estructuralismo*. Buenos Aires: Calder.

Gonseth, F. 1948. "Remarque sur l'idée de complémentarité." *Dialectica* 2, nos. 3–4: 413–20.

Gottmann, Jean. 1947. "De la méthode d'analyse en géographie humaine." *Annales de Géographie* 56, no. 301: 1–12.

Gottmann, Jean. 1952. *La politique des états et leur géographie*. Collection Sciences Politiques. Paris: A. Colin.

Gould, Peter. 1964. "A Note on Research into the Diffusion of Development." *Journal of Modern African Studies* 2, no. 1: 123–25.

Gould, Peter. 1969. "Methodological Developments since the Fifties." *Progress in Geography* 1, no. 20: 1–49.

Gourou, Pierre. 1973. *Pour une géographie humaine*. Nouvelle Bibliothèque Scientifique. Paris: Flammarion.

Gramsci, Antonio. 1999. *A Gramsci Reader: Selected Writings, 1916–1935*. London: Lawrence and Wishart.

Graves, Norman John, and Terence Moore. 1972. "The Nature of Geographical Knowledge." In *New Movements in the Study and Teaching of Geography.* Ed. Norman John Graves, 17–28. London: Temple Smith.

*Grigg, David. 1970. "Region, Models and Classes." In *Integrated Models in Geography: Part IV of Models in Geography.* Ed. Richard J. Chorley. London: Methuen.

Grisoni, Dominque, and Robert Maggiori. 1975. "L'actualisation de l'utopie." *Les Temps Modernes* 30, no. 343: 879–928.

Gurvitch, Georges. 1968. "La vocation actuelle de la sociologie." In *Vers la sociologie differentielle.* Paris: Presses Universitaires de France.

Haeckel, Ernst Heinrich P. A. 1876. *The History of Creation: Or, the Development of the Earth and Its Inhabitants by the Action of Natural Causes.* Trans. and rev. E. R. Lankester. London: Henry S. King.

*Hägerstrand, Torsten. 1965. "A Monte Carlo Approach to Diffusion." *European Journal of Sociology* 6, no. 1: 43–67.

Hägerstrand, Torsten. 1966. "Aspects of the Spatial Structure of Social Communication and the Diffusion of Information." *Papers of the Regional Science Association* 16, no. 1: 27–42.

Hägerstrand, Torsten. 1967a. *Innovation Diffusion as a Spatial Process.* London: University of Chicago Press.

*Hägerstrand, Torsten. 1967b. "On the Monte-Carlo Approach to Diffusion." In *Quantitative Geography: Part 1: Economic and Cultural Topics.* Northwestern Studies in Geography. Ed. W. L. Garrison and D. Marble, 1–32. Evanston, Ill.: Northwestern University Press.

*Hägerstrand, Torsten. 1973. "The Domain of Human Geography." In *Directions in Geography.* Ed. Richard J. Chorley, 67–87. London: Methuen.

Hägerstrand, Torsten. 1974. "Commentary on Anne Buttimer's *Values in Geography.*" Commission on College Geography Resource Paper no. 24. Washington, D.C.: Association of American Geographers.

Haggett, Peter. 1965. *Locational Analysis in Human Geography.* London: Edward Arnold.

Haggett, Peter, and Richard J. Chorley. 1965. "Frontier Movements and the Geographical Tradition." In *Frontiers in Geographical Teaching.* Ed. Richard J. Chorley and Peter Haggett. London: Methuen.

Hall, A. D., and Fagen, R. E. 1968. "Definition of System." In *General Systems Theory.* Vol. 1. Ed. Ludwig Von Bertalanffy. New York: George Braziller.

Harnecker, Marta. 1973. *Los conceptos elementales del materialismo histórico.* Buenos Aires: Siglo XXI.

Hartshorne, Richard. 1939. "The Nature of Geography." *Annals of the Association of American Geographers* 29: 173–658.

Harvey, David. 1967. "Models of the Evolution of Spatial Patterns in Human Geography." In *Integrated Models in Geography.* Ed. Richard J. Chorley and Peter Haggett. London: Methuen.

Harvey, David. 1969. *Explanation in Geography*. London: Edward Arnold.

Harvey, David. 1972. "The Role of Theory." In *The Changing Field of Geography*. Ed. N. J. Graves, 29–41. London: Temple Smith.

Harvey, David. 1973. *Social Justice and the City*. London: Edward Arnold.

Haudricourt, André. 1964. "La technologie, science humaine." *La Pensée* 115: 28–35.

Haushofer, Karl. 1944. *Geopolítica*. México: Fondo de Cultura Económica.

Haveman, Robert H. 1971. *Dialectica sin dogma; ciencia natural y concepción del mundo*. Barcelona: Ariel.

Hegel, Georg Wilhelm Friedrich. 1955. *Einleitung: Die Vernunft in der Geschichte*. Leipzig: Meiner.

Hegel, Georg Wilhelm Friedrich. 1962. *Philosophy of Right*. Trans. and ed. Thomas Knox. London: Oxford University Press.

Hegel, Georg Wilhelm Friedrich. 1966. *Texts and Commentary: Hegel's Preface to His System in a New Translation with Commentary on Facing Pages*. Trans. and ed. Walter Kaufmann. New York: Anchor Books

*Hegel, Georg Wilhelm Friedrich. 1974. *Enciclopedia de las ciencias filosóficas*. Ed. Joan Pablos. México, D.F.: Grijalbo.

*Hettner, Alfred. 1905. "Das Wesen und die Methoden der Geographie." *Geographische Zeitschrift* 11, no. 10: 545–64.

*Hettner, Alfred. 1929. "Unsere Auffassung von der Geographie." *Geographische Zeitschrift* 35, no. 7: 486–91.

Hicks, John. 1959. *Essays in World Economics*. Oxford: Clarendon Press.

Hla Myint, U. 1965. *The Economics of the Developing Countries*. Hutchinson University Library: Economics. London: Hutchinson.

Hobsbawm, Eric. 1964. "Introduction to Marx." In *Pre-Capitalist Economic Formations*. By Karl Marx and Frederick Engels. London: Lawrence and Wishart.

Hodder, B. W. and R. Lee. 1974. *Economic Geography*. London: Methuen.

Hudson, John. 1969. "Diffusion in a Central Place System." *Geographical Analysis* 1, no. 1: 45–58.

Hurst, Michael E. Eliot. 1973. "Establishment Geography: Or How to Be Irrelevant in Three Easy Lessons." *Antipode* 5, no. 2: 40–59.

Husserl, Edmund. 1975. *La crise de l'humanité européenne et la philosophie*. Paris: La Pensée Sauvage.

Huxley, Julian. 1963. "The Future of Man: Evolutionary Aspects." In *Man and His Future*. Ed. Gordon Wolstenholme. London: Little, Brown.

Inkeles, Alex. 1975. "The Emerging Social Structure of the World." *World Politics* 27, no. 4: 467–95.

Isard, Walter. 1956. *Location and Space-Economy: A General Theory Relating to Industrial Location, Market Areas, Land Use, Trade, and Urban Structure*. Regional Science Studies Series (Aldershot, England) 1. Cambridge, Mass.; London: MIT Press.

Isard, Walter. 1960. *Methods of Regional Analysis: An Introduction to Regional Science.* Cambridge, Mass: MIT Press.

Jakobson, Leo, and Ved Prakash. 1971. "Urbanization and Urban Development: Proposals for an Integrated Policy." In *Urbanization and National Development, South and South-East Asian Urban Affairs.* Vol. 1. Beverly Hills, Calif.: Sage.

Jakubowski, Franz. 1971. *Les superstructures idéologiques dans la conception matérialiste de l'histoire.* Paris: Editions de l'Atelier.

Jammer, Max. 1954. *Concepts of Space: The History of Theories of Space in Physics.* Cambridge, Mass.: Harvard University Press.

Jefferson, Mark. 1939. "The Law of the Primate City." *Geographical Review* 29, no. 2: 226–32.

Johnson, E. A. J. 1970. *The Organization of Space in Developing Countries.* Cambridge, Mass.: Harvard University Press.

Jones, Emrys. 1966. *Towns and Cities.* London: Oxford University Press.

Jordan, Z. A. 1971. "Introduction." *Karl Marx: Economy, Class and Social Revolution.* London: Michael Joseph.

Kalesnik, S. V. 1971. "On the Significance of Lenin's Ideas for Soviet Geography." *Soviet Geography* 12, no. 4: 196–205.

Kant, Immanuel. 1802. *Physische Geographie.* Königsberg: F. T. Rist.

Kant, Immanuel. 1929. *Immanuel Kant's Critique of Pure Reason.* London: Macmillan.

Kaufman, Walter. 1966. *Hegel: Texts and Commentary.* New York: Anchor Books.

Kayser, Bernard. 1966. "Les divisions de l'éspace géographique dans les pays sous-développés." *Annales de Géographie* 75: 686–97.

Kelle, V., and M. Kovalson. 1973. *Historical Materialism: An Outline of Marxist Theory of Society.* Moscow: Progress.

Kerblay, Basile. 1966. "A. V. Chayanov: Life, Career, Works." In A. V. Chayanov, *The Theory of Peasant Economy.* Ed. Daniel Thorner, Basile Kerblay, and R. E. F. Smith. Manchester: Manchester University Press.

Keynes, John Maynard. 1939. "The League of Nations Professor Tinbergen's Method." *Economic Journal* 49, no. 195: 558–77.

Klir, Jiri. 1966. "The General Systems as a Methodological Tool." *General Systems* X: 29–42.

Kopnin, Pavel Vasilevich. 1969. *Hipótesis y verdad.* Trans. Lydia Kuper de Velasco. México: Grijalbo.

Kosík, Karel. 1963. *Dialéctica de lo concreto.* México: Grijalbo.

Kuhn, Thomas S. 1962. *The Structure of Scientific Revolutions.* Chicago: University of Chicago Press.

Kusmin, Usivolod. 1974. "Systemic Quality." *Social Sciences* 4: 64–77.

Labriola, Antonio. 1902. *Essais sur la conception matérialiste de l'histoire.* 2nd éd. Trad. Alfred Bonnet. Paris: V. Giard et E. Brière.

Lacoste, Yves. 1976. "Enquête sur le bombardement des digues du fleuve Rouge (Vietnam, été 1972): méthode d'analyse et réflexions d'ensemble." *Hérodote: Revue de Géographie et de Géopolitique*, no. 1: 86–115.

La Grassa, Gianfranco. 1972. "Modo di produzione, rapporti di produzione e formazione economico-sociale." *Critica Marxista* 10, no. 4: 54–83.

Ledrut, Raymond. 1973. *Sociologie urbaine*. Paris: Presses Universitaires de France.

Lee, Alec M. 1970. *Systems Analysis Frameworks*. London: Macmillan.

Lefebvre, Henri. 1973. *Espace et politique: le droit à la ville II*. Paris: Anthropos.

Lefebvre, Henri. 1991. *The Production of Space*. Trans. Donald Nicholson-Smith. Oxford: Blackwell.

Lefebvre, Henri. 1996. *Writings on Cities*. Oxford: Blackwell.

*Leroi-Gourhan, André. 1943. "Ethnologie et géographie." *Revue de Géographie Humaine et Ethnologie* 1, no. 1 (January–March): 14–19.

Leslie, Paul. 1961. *Persons and Perception*. London: Faber and Faber.

Lévy, Jacques. 1975. "Pour une géographie scientifique." *Espace Temps* 1, no. 1: 53–65.

Lukács, György. 1960. *Histoire et conscience de classe*. Paris: Minuit.

Lukermann, F. 1964. "Geography as a Formal Intellectual Discipline and the Way in Which It Contributes to Human Knowledge." *Canadian Geographer* 8, no. 4: 167–72.

Mabogunje, Akin L. 1970. "Systems Approach to a Theory of Rural-Urban Migration." *Geographical Analysis* 2, no. 1: 1–18.

Mabogunje, Akin L. 1975. "Geography and the Problems of the Third World." *International Social Science Journal* 27, no. 2: 288–302.

Mach, Ernst. 1906. *Space and Geometry in the Light of Physiological, Psychological and Physical Inquiry*. La Salle, Ill.: Open Court.

Mach, Ernst. 1914. *The Analysis of Sensations: And the Relation of the Physical to the Psychical*. Chicago: Open Court.

Mackinder, Halford. 1943. "The Round World and the Winning of the Peace." *Foreign Affairs* 2, no. 4 (July): 595–605.

Mandel, Ernest. 1975. *Late Capitalism*. London: New Left.

Marchand, Bernard. 1972. "L'usage des statistiques en géographie." *L'Espace Géographique* 1, no. 2: 79–100.

Markham, Clements. 1905. "The Sphere and Uses of Geography." *The Geographical Journal* 26, no. 6: 593–604.

Marrama, Vittorio. 1961. *Política económica de los paises subesarollados*. Madrid: Aguilar.

Martonne, Emmanuel de. 1957. *Traité de geographie physique*. Vol. 1. 9th ed. Paris: A. Colin.

Marx, Karl. 1973. *Diferencia entre la filosofia de la naturaleza según democrito y según epicuro*. Trans. Juan David Garcia. Caracas: Universidad Central de Venezuela.

*Marx, Karl. 1974. *El capital: Libro I, Capítulo VI, Inédito.* Buenos Aires: Siglo XXI.

Marx, Karl, and Frederick Engels. 1947. *The German Ideology.* New York: International.

Marx, Karl, and Frederick Engels.1964a. *Lettres sur Le Capital.* Paris: Sociales.

Marx, Karl, and Frederick Engels. 1964b. *Pre-Capitalist Economic Formations.* Intro. Eric Hobsbawm. London: Lawrence and Wishart.

Marx, Karl, and Frederick Engels. 1969. *Karl Marx and Frederick Engels: Selected Works.* Moscow: Progress.

Mathieu, Nicole. 1974. "Propos critiques sur l'urbanisation des campagnes." *Espaces et Sociétés* 12: 71–89.

May, J. A. 1970. *Kant's Concept of Geography and Its Relation to Recent Geographical Thought.* Toronto: University of Toronto Press.

McCall, Daniel. 1962. "The Koforidua Market." In *Markets in Africa.* Ed. George Dalton and Paul Bohannan, 667–97. Evanston, Ill: Northwestern University Press.

McNulty, Michael L. 1969. "Urban Structure and Development: The Urban System of Ghana." *Journal of Developing Areas* 3, no. 2: 159–76.

Mehedinti, S. 1901. "Le géographie comparée." *Annales de Géographie* 10, no. 49: 1–9.

Meier, Richard L. 1962. *A Communication Theory of Urban Growth.* Joint Center for Urban Studies of the Massachusetts Institute of Technology and Harvard University. Cambridge, Mass.: MIT Press.

Meliujin, Serafin. 1963. *Dialéctica del desarollo en la naturaleza inorgánica.* México: Juan Grijalbo.

*Merleau-Ponty, Maurice. 1964. *Le visible et l'invisible: suivi de notes de travail.* Paris: Gallimard.

*Merleau-Ponty, Maurice. 1971. *Existence et dialectique.* Paris: Presses Universitaires de France.

Michotte, P. 1921. "L'orientation nouvelle en géographie." *Bulletin de La Société Belge de Géographie* 4, no. 1: 5–43.

Miller, George, Eugene Galanter, and Karl Pribram. 1960. *Plans and the Structure of Behavior.* New York: Holt, Rinehart and Wilson.

Montesquieu, Charles. [1748] 1951. *L'ésprit des lois.* Paris: La Pléiade.

Moodie, D. 1971. "Content Analysis: A Method for Historical Geography." *Area* 3, no. 3: 146–49.

Moore, Wilbert E. 1963. *Man, Time, and Society.* London: Wiley.

Morazé, Charles. 1974. "L'histoire, science naturelle." *Annales: Histoire, Sciences Sociales* 29, no. 1: 107–37.

Morgenstern, Irvin. 1960. *The Dimensional Structure of Time: Together with the Drama and Its Timing.* New York: Philosophical Library.

Morrill, Richard L. 1968. "Waves of Spatial Diffusion." *Journal of Regional Science* 8, no. 1: 1–18.

Moya, Carlos. 1970. *Sociologos y Sociologia*. Madrid: Siglo Veintiuno de España.

Novack, George Edward. 1969. *An Introduction to the Logic of Marxism*. New York: Merit.

Oliveira, Livia de. 1977. "Contribuição dos estudos cognitivos à percepção geográfica." *Geografia* 2, no. 3: 61–72.

Olsson, Gunnar. 1967. "Central Place Systems, Spatial Interaction, and Stochastic Processes." *Papers in Regional Science* 18, no. 1: 13–45.

Ortega y Gasset, José. 1963. "History as a System." In *Philosophy and History*. Ed. Raymond Klibansky and H. J. Paton, 283–322. New York: Harper/Torch Books.

Pahl, R. E. 1965. 'Trends in Social Geography'. In *Frontiers in Geographical Teaching*. Ed. Richard J. Chorley and Peter Haggett, 81–100. London: Routledge.

Pailhé, Joël, Christian Grataloup, and Jacques Lévy. 1977. "Texte-débat. 'La géographie procès sans sujet' de Joël Pailhé." *Espace Temps* 5, no. 1: 33–55.

Paivio, Allan. 1975. "Neomentalism." *Canadian Journal of Psychology/Revue canadienne de psychologie* 29, no. 4: 263–91.

Park, Robert E., and E. W. Burgess. 1921. *Introduction to the Science of Sociology*. Chicago: University of Chicago Press.

Parsons, Talcott, and Neil Smelser. 1956. *Economy and Society: A Study in the Integration of Economic and Social Theory*. Glencoe, Ill.: Free Press.

*Pierson, Donald. 1970. "Estudos de ecologia humana." São Paulo: Martins.

Pinto, Aníbal, and Osvaldo Sunkel. 1966. "Latin American Economists in the United States." *Economic Development and Cultural Change* 15, no. 1: 79–86.

Plekhanov, Georgi. 1967. *Materialismo militante*. México: Grijalbo.

Plekhanov, Georgi. 1972. *Fundamental Problems of Marxism*. Moscow: Progress.

Plekhanov, Georgi. 1974. *Oeuvres philosophiques*. Moscow: Progress.

Poincaré, Henri. 1905. *La valeur de la science*. Bibliothèque de philosophie scientifique. Paris: Flammarion.

Poincaré, Henri. 1914. *Science and Method*. London: T. Nelson.

Pokhishevskiy. V. V. 1975. "Social Geography Problems in the Regulation of Settlement Systems in a Developed Socialist Country." *Soviet Geography* 16, no. 1: 28–40.

Pred, Allan. 1967. *Behavior and Location: Foundations for a Geographic and Dynamic Location Theory*. Lund Studies in Geography. Series B, Human Geography nos. 27–28. Lund: Gleerup.

Prestipino, Giuseppe. 1977. "El pensamiento filosófico de Engels." *Naturaleza y Sociedad En La Perspectiva Teórica Marxista*. México: Siglo XXI.

Randle, Patricio H. 1966. *Geografía Histórica y planeamiento*. Buenos Aires: Eudeba.

Rattner, Henrique. 1964. "Contrastes regionais no desenvolvimento econômico brasileiro." *Revista de Administração de Empresas* 4, no. 11: 133–66.

Rattner, Heinrich. 1972a. "L'espace et la diffusion de l'innovation: les élites 'industrialisantes' en Amérique latine." *Revue Tiers Monde* (October 1): 851–62.

Rattner, Henrique. 1972b. *Industrialização e concentração econômica em São Paulo.* Rio de Janeiro: Fundação Getúlio Vargas.

Ratzel, Friedrich. 1887. *Völkerkunde.* Leipzig: Verlag des Bibliographischen Instituts.

Ratzel, Friedrich. 1896. *History of Mankind.* Trans. A. J. Butler. 3 vols. London: Macmillan.

Ratzel, Friedrich. 1899. *Anthropogeographie.* 2. Aufl. Bibliothek geographischer Handbücher. Stuttgart: J. Engelhorn.

Reclus, Élisée. 1877. *Nouvelle géographie universelle.* Paris: Hachette.

Reichenbach, Hans. 1965. *The Theory of Relativity and A Priori Knowledge.* Berkeley: University of California Press.

Rémy, Jean. 1966. *La ville, phénomène économique.* Bruxelles: Vie Ouvriere.

Resources for the Future. 1966. *Design for a Worldwide Study of Regional Development: A Report to the United Nations on a Proposed Research-Training Program.* Resources for the Future Staff Study. Washington, D.C.; Baltimore: Resources for the Future.

Rey, Pierre-Philippe. 1973. *Les alliances de classes.* Paris: Maspero.

Ribeiro, Darcy. 1969. "Las Américas y la civilización." In *La civilización occidental y nosotros: los pueblos terstimonio.* Buenos Aires: Centro Editor de América Latina.

Ricci, François. 1974. "Structure logique du paragraphe I du Capital." In *Logique de Marx.* Ed. Jacques D'Hondt, 105–33. Paris: Presses Universitaires de France.

Richardson, Harry W. 1969. *Regional Economics: Location Theory, Urban Structure and Regional Change.* London: Weidenfeld and Nicolson.

Riddell, J. Barry. 1970. "On Structuring a Migration Model." *Geographical Analysis* 2, no. 4: 403–9.

Riesner, Richard. 1973. "The Territorial Illusion and Behavioral Sink: Critical Notes on Behavioural Geography." Antipode 5, no. 3: 52–57.

Rimbert, Sylvie. 1972. "Aperçu sur la géographie théorique. Une philosophie, des méthodes, des techniques." *L'Éspace Géographique* 1, no. 2: 101–6.

Ritter, Carl. 1863. *Geographical Studies.* Boston: Gould and Lincoln.

Ritter, O. 1974. "La configuration des continents sur la surface de la terre." In *Introduction à la géographie générale comparée,* 217–41. Paris: Les Belles Lettres.

Rodoman, B. B. 1973. "Territorial Systems." *Soviet Geography* 14, no. 2: 100–105.

Rofman, Alejandro, and L. A. Romero. 1974. *Sistema socio-económico y estructura regional en la Argentina.* Buenos Aires: Amorrurtu.

Rofman, Alejandro B. 1974a. *Dependencia, estructura de poder y formación regional en América Latina.* 1st ed. Sociología y política. Buenos Aires: Siglo Veintiuno Editores.

Rofman, Alejandro B. 1974b. *Desigualdades regionales y concentración económica: el caso argentino.* 1st ed. en español. Buenos Aires: Ediciones Siap-Planteos.

*Ruellan, Francis. 1943. "As normas de elaboração e redação de um trabalho geográfico." *Revista Brasileira de Geografia* 5, no. 4.

Russell, Bertrand. 1945. *A History of Western Philosophy*. New York: Simon and Schuster.

Russell, Bertrand. 1958. *The ABC of Relativity*. London: Kegan Paul, Trench, Trubner.

Russell, Bertrand. 1966. *Human Knowledge: Its Scope and Limits*. Muirhead Library of Philosophy. New York: Allen and Unwin.

Santayana, George. 1924. *Scepticism and Animal Faith: Introduction to a System of Philosophy*. London: Constable and Co.

Santos, Milton. 1967. *Croissance démographique et consommation alimentaire dans les pays sous-développés*. Paris: Centre de Documentation Universitaire.

Santos, Milton. 1968. "La géographie urbaine et l'économie des villes dans les pays sousdéveloppés." *Revue de Géographie de Lyon* 43, no. 4: 362–76.

*Santos, Milton. 1970. *Dix essais sur les villes des pays sous-développés*. Paris: Ophrys.

Santos, Milton. 1971a. *Le métier de géographe en pays sous-développé*. Paris: Ophrys.

Santos, Milton.1971b. "Analyse régionale et aménagement de l'espace." *Revue Tiers Monde* 12, no. 4: 199–203.

Santos, Milton. 1972. "Dimension temporelle et systèmes spatiaux dans les pays du Tiers Monde." *Tiers Monde* 13, no. 50: 247–68.

Santos, Milton. 1974a. "Geography, Marxism and Underdevelopment." *Antipode* 6, no. 3: 1–9.

Santos, Milton. 1974b. "Sous-développement et pôles de croissance économique et sociale." *Revue Tiers Monde* 15, no. 5: 271–86.

Santos, Milton. 1974c. "Time-Space Relations in the Underdeveloped World." University of Dar-es-Salaam, Department of Geography.

Santos, Milton. 1975a. *L'espace partagé: les deux circuits de l'économie urbaine des pays sous-développés*. Paris: Génin.

Santos, Milton. 1975b. "Space and Domination—A Marxist Approach." *International Social Science Journal* 27, no. 2: 346–63.

Santos, Milton. 1976. "Relações espaço-temporais no mundo subdesenvolvido." *Seleção de Textos* 1. Associação dos Geógrafos Brasileiros, Secção Regional de São Paulo, Dezembro.

Santos, Milton. 1977a. "Sociedad y espacio: la formación social como teoría y como método." *Revista Latino Americano de Economia*, México.

Santos, Milton. 1977b. "Sociedade e espaço: a formação social como teoria e como método." *Boletim Paulista de Geografia*, no. 54.

Santos, Milton. 1977c. "Société et éspace: la formation sociale comme théorie et comme méthode." *Cahiers Internationaux de Sociologie*. Nouvelle série, 63 (juillet–décembre): 261–76.

Santos, Milton. 1977d. "Society and Space: Social Formation as Theory and Method." Trans. Stephan Slaner. *Antipode* 9, no. 1: 3–13.

Santos, Milton. 1977e. "Réponses à Michel Foucault." In O. Bernard and M. Ronai, "Des réponses aux questions de Michel Foucault." *Hérodote* 6: 5–30.

Santos, Milton. 1978a. "Sociedad y espacio: la formación social como teoría y como método." *Cuadernos Venezolanos de Planificación.*

Santos, Milton. 1978b. *The Shared Space: The Two Circuits of Urban Economy in Underdeveloped Countries and Their Spatial Repercussions.* London: Methuen.

Santos, Milton. 1978c. *Economia espacial: críticas e alternativas.* São Paulo: Hucitec.

Sartre, Jean-Paul. 2004. *Critique of Dialectical Reason.* Trans. Alan Sheridan-Smith. London: Verso.

Sauer, Carl. 1931. *An Introduction to Geography. I: Elements.* Ann Arbor, Mich.: Edwards Bros.

Sauer, Carl. 1962. "Cultural Geography." In *Readings in Cultural Geography.* Ed. Philip L. Wagner and Marvin W. Mikesell. Chicago: University of Chicago Press.

Sauer, Carl Ortwin. 1963. *Land and Life: A Selection from the Writings of Carl Ortwin Sauer.* Berkeley: University of California Press.

Sautter, Gilles. 1974. "Crise ou renouveau de la géographie." *Travaux de l'Institut de Géographie de Reims* 20, no. 1: 101–8.

Sautter, Gilles. 1975. "Some Thoughts on Geography in 1975." *International Social Science Journal* 27, no. 2: 231–249.

Schilling, H. n.d. "An Operational View." *American Scientist* 52: 388a–96.

Schmidt, Alfred. 1971. *The Concept of Nature in Marx.* London: NLB.

Schön, Donald A. 1973. *Beyond the Stable State: Public and Private Learning in a Changing Society.* London: Maurice Temple Smith.

Schumpeter, Joseph. 1964. *Síntesis de la evolución de la ciencia económica y sus métodos.* Vilassar de Mar, Barcelona: Oikos-tau.

Schumpeter, Joseph. 1970. *Capitalism, Socialism, and Democracy.* London: University Books.

Sebag, Lucien. 1972. *Marxismo y structuralismo.* Madrid: Siglo XXI.

Secchi, Bernardo. 1968. "Bases teóricas del análisis territorial." In *Análisis del Las Estructurales Territoriales,* 17–99. Barcelona: Gustavo Gili.

Sethuraman, S. V. 1974. "Towards a Definition of the Informal Sector." Mimeo.

Shonfield, Andrew. 1969. "Thinking about the Future." *Encounter* 32: 15–26.

*Silva, Armando Corrêa da. 1972. "Ciência e valor em geografia." In *Métodos em Questão No. 4.* Publicações Do Instituto de Geografia. São Paulo: Universidade de São Paulo.

Silva, Armando Corrêa da. 1976a. "Geografia e ideologia." *Boletim Paulista de Geografia* 52: 93–100.

Silva, Armando Corrêa da. 1976b. "Uma proposição teórico em geografia." In *Metodos em Questão No. 13.* Publicações Do Instituto de Geografia. São Paulo: Universidade de São Paulo.

*Slater, David. 1975a. "The Poverty of Modern Geographical Enquiry." *Pacific Viewpoint.* 16, no. 2: 159–76.

*Slater, David. 1975b. "Underdevelopment and Spatial Inequality: Approaches to the Problems of Regional Planning in the Third World." *Progress in Planning* 4: 97–167.

Smailes, Arthur E. 1953. *The Geography of Towns.* London: Hutchinson.

Sodré, Nelson Werneck. 1976. *Introdução à geografia: geografia e ideologia.* Petrópolis: Vozes.

Sorre, Max. 1948. "The Concept of Genre de Vie." In *Readings in Cultural Geography.* Ed. Philip L. Wagner and Marvin W. Mikesell, 399–415. Chicago: University of Chicago Press.

Sorre, Max. 1952. *Introduction. Tome II, L'Habitat: Les fondements de la géographie humaine.* Paris: Colin.

Sorre, Max. 1957. *Recontres de la géographie et de la sociologie.* Paris: Marcel Rivière.

Sorre, Max. [1953]1962. "The Role of Historical Explanation in Geography." In *Readings in Cultural Geography.* Ed. Philip L. Wagner and Marvin W. Mikesell, 44–47. Chicago: University of Chicago Press.

Sportelli, Silvano. 1974. "A proposito della teoria Sartriana del practico-inerte." *Critica Marxista* 12, no. 5: 77–97.

Staats, Arthur. 1975. *Social Behaviourism.* Belmont, Calif.: Dorsey Press.

Stoddart, D. 1967. "Organism and Ecosystem as Geographical Models." In *Integrated Models in Geography.* Ed. Richard Chorley and Peter Haggett. London: Methuen.

Taaffe, Edward J., Richard L. Morrill, and Peter R. Gould. 1973. "Transport Expansion in Underdeveloped Countries: A Comparative Analysis." In *Transport and Development.* Ed. B. S. Hoyle, 32–49. New York: Springer.

Taylor, Griffith. 1946. *Our Evolving Civilization: An Introduction to Geopacifics, Geographical Aspects of the Path toward World Peace.* Toronto: University of Toronto Press.

*Taylor, Griffith. 1951. *Geography in the Twentieth Century.* London: Methuen.

Theodorsen, George. 1961. *Studies in Human Ecology.* New York: Harper and Row.

Thompson, D'Arcy Wentworth. 1917. *On Growth and Form.* Cambridge: Cambridge University Press.

*Thünen, Johann Heinrich von. 1826. *Der isolierte Staat in Beziehung auf Landwirtschaft und Nationalökonomie.* Hamburg.

Tovar, Ramón A. 1974. *Lo geográfico.* Caracas: Instituto Pedagógico.

Tucey, Mary L. 1976. "Cognitive-Behavioural Approaches in Geography; the Search of a New Model of Man." Discussion paper, Department of Geography, University of Dar-es-Salaam.

Tulippe, Omer. 1945. *Cours de géographie humaine: géographie humaine générale. Tome 1.* Paris: Desoer.

*Ullman, Edward L. 1953. "Human Geography and Area Research." *Annals of the Association of American Geographers* 43, no. 1: 54–66.

Ullman, Edward L. 1973. "Ecology and Spatial Analysis: A Comment on James D. Clarkson." *Annals of the Association of American Geographers* 63, no. 2: 272–74.

Vagaggini, V. Y., and G. Dematteis. 1977. "El método analítico de la geografía." *Revista Terra*, no. 1.

Vidal de La Blache, P. 1896. "Le principe de la géographie générale." *Annales de Géographie* 5: 129–42.

Vidal de La Blache, P. 1899. "Leçon d'ouverture du cours de géographie: Faculté des Lettres de Paris, 7 février 1899." *Annales De Géographie* 8, no. 38: 97–109.

Vidal de La Blache, P. 1911. *Les genres de vie dans la géographie humaine*. Paris: A. Colin.

Vieille, Paul. 1974. "L'éspace global du capitalisme d'organisation." *Éspaces et Sociétés* 12: 3–32.

Ville de Bordeaux. 1948. *Montesquieu et l'esprit des lois, 1748–1948*.

*Von Bertalanffy, Ludwig. 1950. "An Outline of General System Theory." *British Journal for the Philosophy of Science* 1: 134–65.

*Von Bertalanffy, Ludwig. 1962. "General Systems Theory: A Critical Review." *General Systems* 7: 1–20.

Von Bertalanffy, Ludwig. 1968. *General Systems Theory*. New York: George Braziller.

Wagemann, Ernst. 1933. *Estructura y ritmo de la economia mundial*. Barcelona: Labor.

Wagner, Philip L., and Marvin W. Mikesell. 1962a. "Introduction." In *Readings in Cultural Geography*. Ed. Philip L. Wagner and Marvin W. Mikesell. Chicago: University of Chicago Press.

Wallman, Sandra. 1975. "Kinship, a-Kinship, Anti-Kinship: Variation in the Logic of Kinship Situations." *Journal of Human Evolution* 4, no. 5: 331–41.

Watson, J. W. 1951. "The Sociological Aspects of Geography." In *Geography in the Twentieth Century*. Ed. Griffith Taylor. London: Methuen.

Weiss, Paul. 1958. *Modes of Being*. Carbondale: Southern Illinois University Press.

Wettstein, German. 1973. "Una geografía de los paises dependientes." *Ciencianueva 25*. Montevidéo.

*Wettstein, German, José Rojas Lopes, and Jovito Valbuena. 1976. *La percepción en geografía*. Escuela de Geografía, Cuadernos 49. Merida, Venezuela: Universidade de los Andes.

Whitehead, Alfred North. 1929. *Process and Reality: An Essay in Cosmology*. Gifford Lectures 1927. Cambridge: Cambridge University Press.

Whitehead, Alfred North. 1938. *Modes of Thought*. New York: Macmillan.

Whitehead, Alfred North. 1948. *Essays in Science and Philosophy*. London: Rider.

Whitehead, Alfred North. 1964. *The Concept of Nature*. Cambridge: Cambridge University Press.

Whittlesey, Darwent. 1957. "The Regional Concept and Regional Method." In *American Geography: Inventory and Prospect*. Ed. Preston James and C. F. Jones. Syracuse: Syracuse University Press.

*Wilson, Alan G. 1967. "A Statistical Theory of Spatial Distribution Models." *Transportation Research* 1: 253–69.

Wilson, Alan G. 1969. "The Use of Analogies in Geography." *Geographical Analysis* 1, no. 3: 225–33.

Wittgenstein, Ludwig. 1969. *Tractatus Logico-Philosophicus.* International Library of Psychology, Philosophy, and Scientific Method. London: Routledge and Kegan Paul.

*Wolpert, Julian. 1966. *A Regional Simulation Model of Information Diffusion.* Philadelphia. Mimeo.

Woodbridge, Frederick J. E. 1940. *An Essay on Nature.* New York: Columbia University Press.

Wrigley, E. A. 1965. "Changes in the Philosophy of Geography." In *Frontiers in Geographical Teaching.* Ed. Richard J. Chorley and Peter Haggett, 17. London: Methuen.

Index